The Little Book of Autorotations

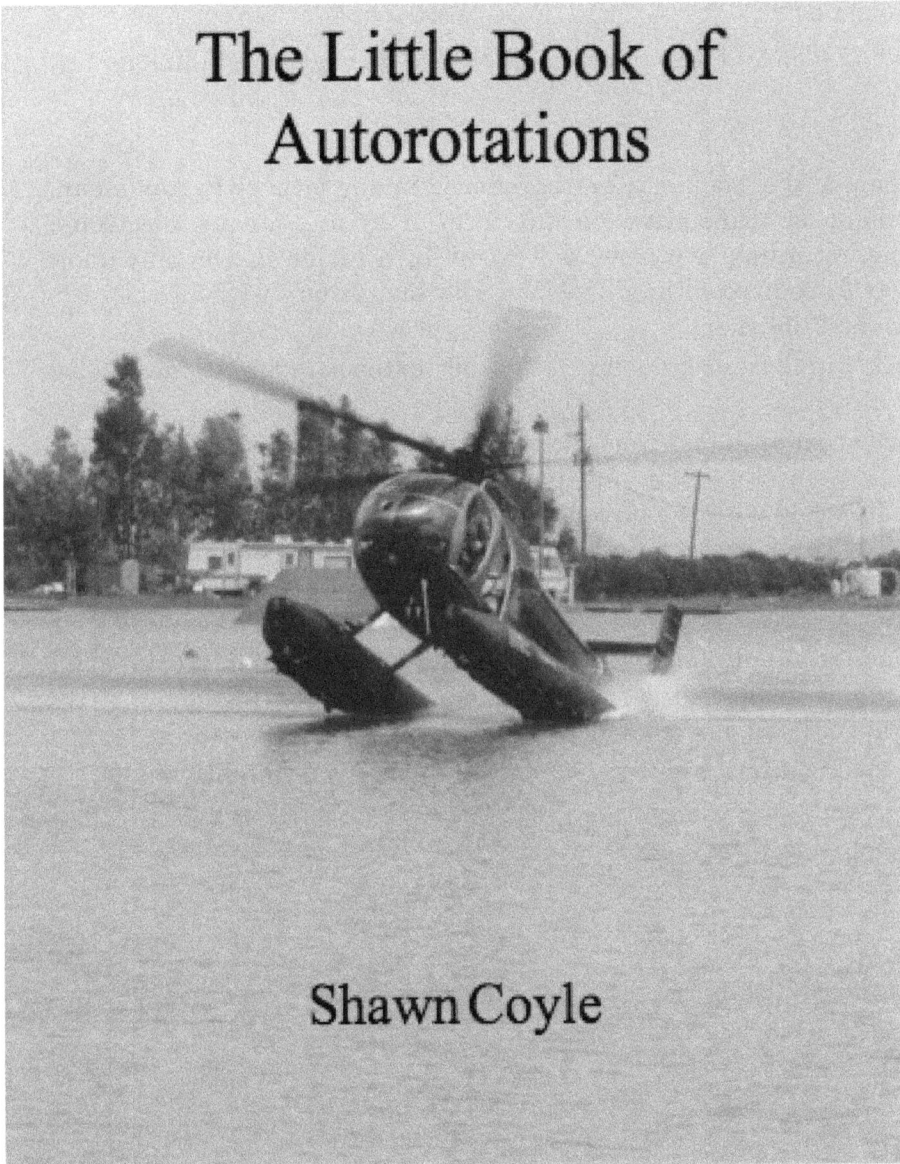

Shawn Coyle

The Little Book of Autorotations

For information on bulk purchases, academic sales or textbook adaptations, please contact Eagle Eye Solutions, LLC.

This book contains information gathered from many sources. It is published for general reference and not as a substitute for independent verification by users when circumstances warrant. It is sold with the understanding that the author is not engaged in rendering any legal advice or explicit flight instruction. The publisher and author disclaim any personal liability, either directly or indirectly for advice or information presented within.

Every effort has been made to supply complete and accurate information, however Eagle Eye Solutions, LLC assumes no liability for its use, nor for any infringement of the intellectual property rights of third parties which would result from such use.

Any slights of persons, organizations, publishers, books or places is unintentional.

Library of Congress Card Number

ISBN 9780979263842

Manufactured in the United States.
Published by:
Eagle Eye Solutions, LLC
P.O. Box 237
Lebanon, OH. 45036-0237
Phone 1-800-653 7483 (USA)
email: shawncoyle@earthlink.net
www.EagleEyeSolutionsLLC.com

Publishers Cataloging in Publication Data
Coyle, Shawn C
The Little Book of Autorotations
First Edition
Bibliography: p.
Includes index.
1. Books - United States - Aviation   1. Coyle, Shawn, 1950 -
II Title – The Little Book of Autorotations, 2013

Preface

Dedication

This book is dedicated to all those from whom I learned the subtle things of flying helicopters, whether they knew it or not. This includes both students and instructors I flew with. Many thanks!

To Pete Gillies, Matt Johnson, Itai Oran and others who made suggestions and comments. Many thanks also. The book would have not been the same without your input.

As always this book is also dedicated to the Chief Pilot of the Universe, who guides us through the tricky currents of the world, without or without earthly engine power.

Table of Contents

Chapter 1   Introduction

Helicopter engines, being of earthly construction, are bound to fail. Since a single engine helicopter depends solely upon this device for support, losing it is an inconvenience at a minimum, and at times, critical. You need to understand that the loss of an engine need not be the end of the world.

It is very prudent to know the symptoms of an engine failure, what must be done about this inconvenience, and your options for any given situation.

Statistics show since helicopters stop first and then land, the chances of surviving an engine failure are much better than surviving an engine failure in a fixed wing airplane.

The first part of this book covers the very basics of engine failures and autorotations in single--engine helicopters - it should be sufficient for the beginner. For example, no turns are considered initially. More advanced techniques are covered later.

The "normal" sequence of events that should occur when practicing autorotations will vary from helicopter type to type; but the principles remain the same. It must be stressed autorotations are not a "by the numbers" procedure. Even though a lot of the current training methods appear to make it look like that.

Some normal disclaimers are also needed - first, all the techniques here are generic, and may not always work on a particular type of helicopter. Secondly, the aim of teaching autorotations must be firmly remembered - there is no point bending the helicopter to split hairs about esoteric points of technique - if the helicopter can be maneuvered safely to a low height above ground and low groundspeed, the aim has been achieved. The following impact should be survivable.

Each of the areas will be covered in detail in separate chapters. There is a lot to cover in this very rapid maneuver, and it is normal for the student to not follow or understand what is happening the first time they see an autorotation.

The first time a highly dynamic maneuver is shown it's normal for the student to utter a "What just happened?" -- it's called time compression and it happens to us all. The first time a student sees a touch down autorotation there's going to be a very large "What was that?" factor. This book attempts to explain the process in small steps, and provide exercises to develop the pilot's ability to use all the possibilities to best advantage.

Terminology

Since helicopters rotors don't all turn the same direction, to avoid confusion, I'll use the term "power pedal" to refer to the pedal that is pushed forward when collective is raised, and "non-power pedal" to refer to the pedal that is pushed forward when the collective is lowered or engine power is reduced. Similarly, the term "chop" will not be applied to moving the throttle to idle, as sometimes too hard or fast a movement can turn a simulated engine failure into a real one. I know…

Some Interesting Concepts

Before we discuss that, though, let's examine a couple of key, but controversial statements. The first is autorotations are:

perhaps the most sensory-deprived, multi-variable maneuver

a pilot can be asked to perform. The second is autorotations:

Is not a by-the-numbers maneuver.

What do I mean by these statements and why were they said?

Sensory-deprived

Sensory deprived means that the pilot doesn't have a lot of readily available, useful information to judge what to do when during the maneuver - things like airspeed, power, height above the ground, and so on. Instruments are there, to be sure, but by the end of this book you should be convinced that they're really not much use to the pilot in an autorotation.

For most things in helicopter flying, we have a lot of information to help complete the task. In an approach with power, for example, we have a definite landing point to aim for, power to manage and a whole host of visual cues to tell us what to do. When flying an instrument approach, we have the various flight and navigation instruments to tell us what's going on and what we have to do.

In a real autorotation, we are suddenly faced with a whole new set of conditions. We no longer have much choice over when we're going to land - in fact the choice of suitable landing places will be small unless we're at a great height above the ground. None of the engine instruments are of value (the engine has stopped) -the rotor RPM gauge might be useful, but if you can hear the rotor RPM (and know how it changes sound), and are above the minimum power-off rotor RPM, that's all you need to know. Now, you have to work out how to get to a particular spot, and that's where the multi-variable aspect comes in.

Time will also be very short, which adds enormously to the pressure, and doesn't help to quell a sense of panic.

Multi-variable

By multi-variable, we mean there are a lot of ways to do things. The Chinese saying of "There are a thousand paths to the top of the mountain" can now be turned upside down to become "There are thousand ways down to the ground". Most of them are safe, but some are better than others, and a few are definitely not OK. (To say the least!)

For the helicopter pilot suddenly faced with an engine failure and looking for somewhere to land, the variables become airspeed, flight path, rotor RPM, and rate of descent (there may be others!!!). How you play one (or more) off against the others to arrive safely on the ground is something no book can tell you. Do you use lots of airspeed and low rotor RPM to get maximum range? What if you're too close and high? What if the only suitable landing spot is a hole in very tall trees?

The statement that, "An autorotation is not a by-the-numbers maneuver," is going to get folks talking. I firmly believe, though, that there is no one set of numbers such as airspeed, flare attitude, and so on that is going to guarantee a safe landing in every situation. By the end of this book, you should also be convinced.

There is little point in always starting training autorotations from the same spot--.

"aligned-with-the-landing-runway-abeam-the-water-tower-at-1,200-feet-60-knots,-level-flight,-lower-the-collective-and-roll-the-throttle-off-on-3-2-1".

The real world isn't that charitable. More on that later.

No Aerodynamics!!!

Some of you will probably be happy that there is not going to be the usual collection of vector diagrams in this book. Others will be ecstatic! For those who are disappointed -- look at Cyclic and Collective!

The reason for leaving out these diagrams is that while they're important for a basic understanding, they're not as important as the knowledge of what you see in and from the cockpit. We're going to concentrate on that, and I don't recall a lot of graphs and vector diagrams suddenly springing up either in the cockpit or getting projected onto the windshield.

Necessarily, there will be a few charts, and only a few simple equations.

Who Should Teach Autorotations??

In a later chapter, we are going to discuss where you should (and shouldn't) do training touchdown autorotations. As important as that is, anyone studying the topic of training autorotations needs to first consider: who is going to teach autorotations?

The Importance of Experience

Why is the experience of the person teaching autorotations important? Let's first define experience. Experience is not just flight hours.

A formal definition of experience might be:

> *Knowledge of, or skill in, or observation of, some thing or event gained through involvement in or exposure to that thing or event.*

Experience thus requires involvement in an event to gain the respective knowledge or skill.

To teach any subject, it would be reasonable to expect the instructor to have more experience in this particular area than the student.

I have flown with pilots who have 7,000 hours, which turned out to be 7,000 repetitions of the same thing, with little (or no apparent) learning involved. Likewise, I've flown with pilots who have only 900 hours, but possessed the Wisdom of Solomon.

Since there are a whole host of variables in any autorotation, worthwhile experience in autorotations must include a variety of different situations.

Experience in autorotations boils down to knowing the variables and how to use them. Experience in teaching autorotations is knowing how to teach the variables in a manner that is progressive, easy to understand and remember.

Experience is also perishable. The degree of loss of experience will depend on both time and depth of experience. It may be a long time since you rode a bicycle, but if you had ridden one a lot, it comes back pretty quickly.

Constructing the Teacher

If you're an instructor, just because you have a lot of flying time, or even a lot of time doing autorotations, doesn't mean you can teach them well. Going through our growing list of requirements, we know the right teacher must have a lot of overall experience, recent experience, and can teach autorotations. That's a pretty steep order, but it's not complete!

A good teacher also needs to be able to teach a variety of autorotation situations, not just the "standard" type. The teacher needs to be able to get the student to make-it-to-that-spot from within the helicopter's capability with a good level of competency, not just to be able to pass the check ride.

A good teacher has the student learn the variables and how to use them. Since we've established that there are no by-the-numbers procedures that work correctly every time, we need to establish how to use the variables. More on all of those variables later.

Experience means knowing the variables to use and when to use which ones. This part of the learning must actually involve the student.

Students will make mistakes and have errors in technique, which is an important part of learning. It is important to recognize errors in technique early and correct for them immediately. The instructor needs to let the student learn to recognize those mistakes and how to correct them, then ask for an explanation of why the teacher did something (after the event, of course). The change in behavior as a result of experience is how we learn!

The discussion could quickly degenerate into the minutiae of how to teach autorotations, but that's not the point of this chapter. If you're an instructor who knows the variables, knows how to teach them and then let the student build experience, so he or she can apply it in everyday flying. Then you're successful. It will take some time and effort.

Also, even instructors need refreshers every once in a while. I recommend the factory training school for the machine you're instructing in.

Sequence of Chapters and Training Exercises

The sequence of the following chapters may seem too odd when first reviewed. The layout is deliberate, in order to present information to the student in a logical manner with a good deal of build-up. Once all the training maneuvers have been completed, then the warm up sequence shown earlier can be used, but it is recommended that the initial training be in the order shown in the next few chapters.

A Training Regimen

For student pilots, the initial experiences should be incremental and the student should demonstrate proficiency at each step. The first parts should not be drawn out, but concentrated as a package, so that the learning points sink in quickly. The build-up in skills will go a long way to success. The training used to introduce the necessary points will be different than that used later to consolidate them all. For example, we start in a long, steady descent from a good height above the ground to let the student understand what happens to rotor RPM in turns and airspeed changes. Later, it's assumed that these points were learned and understood and can be applied without a lot of review.

When we later consolidate the set of skills to cover the whole maneuver, the build up follows progressive steps, detailed in Chapter 8.

The total package should take a maximum of 4 flight hours to accomplish.

It is possible to accomplish nearly all of the learning points without a great number of touchdowns. More on that later.

A moderate number of autorotations done in a short period of time will make for better learning than a large number stretched out over a long period. Keep in mind, though, after about 90 minutes, learning tapers off pretty quickly. Pace both instructor and student.

After initial training, refreshers are needed. Hopefully, every month or so a practice autorotation can be made (even a power recovery is better than nothing). But, again, the learning has to be learning, not just here-we-go-again. It is when we are subjected to different stimuli than normal that we learn and adapt (i.e., gain experience).

Once the student is able to make the helicopter arrive over the preferred spot from anywhere in the matrix of variables that permit, the feeling of satisfaction will be quite high. Getting there is a real thrill, and understanding all the possibilities is definitely a growth experience.

Factory Training Schools / Manufacturer's Schools

Does this sound like I'm advocating using the manufacturer's factory training school? You bet. There are few other places that have instructors who have a lot of experience in teaching autorotations, who have recent experience teaching them and who are all excellent instructors, especially at autorotations. For other places, always check if they have the depth and recency of experience, and the ability to teach how to use all the variables.

Student Debriefing of Autorotations

Having the student debrief each autorotation immediately after it happens develops skills at observing and correcting mistakes, as well as helping them to relax. If possible, have the student talk throughout the maneuver about what is going on-- this helps when the debriefing comes. One instructor insisted the student recite nursery rhymes to improve relaxation. Video recording the autorotation from inside the cockpit helps a lot....

Eyes Out of the Cockpit!

On a philosophical note, if the engine really fails, there is almost nothing inside the cockpit worth looking at to help the pilot land the helicopter successfully. First of all, none of the engine instruments are worth looking at - the engine has failed. Secondly, none of the other instruments help the pilot to judge when, where and how much to move the controls to make it to the selected landing spot. Altitude from a pressure altimeter is of no use - it's just a number, and may be useful only if it is height above ground (but it's never, ever height above the ground[1]). Most light helicopters don't have radar altimeters, and there is no magic formula to tell what height to be at what distance - there is no way to measure the distance anyway. Airspeed only tells you how fast you are traveling through the air, (as opposed to how fast you need to go to make it to the clearing), and so on. It is far better to concentrate on looking outside.

A third, but less important reason for not looking at the instruments is that every instrument in the cockpit doesn't tell the truth -- all instruments have errors in them, and they all have a bit of a lag in response[2].

The pilot should be looking outside the helicopter over 80% of the time during an autorotation. Any glances inside should be of very short duration to look at a known instrument location followed by looking outside again while the information is digested.

---

[1] For those in the UK who advocate the use of QFE, that is only valid at the airfield you're flying to - way too many places where it's not valid.

[2] The only device in the helicopter that tells the truth is the window, and even it has some distortions.

Some supposedly "useful" instruments - the vertical speed indicator is a good example, are of absolutely no use -- would you change anything in your control strategy if it showed a reading higher or lower than normal at the instant you looked at it?

About the only instrument in the cockpit you might want to look at is the slip ball, and you should be able to tell what's going on in that department almost instinctively. If it shows the helicopter slightly out of trim, is it worth spending a lot of time and effort fussing to put the machine back in perfect balance? Only you can decide at the moment.

First Troubleshooting Tip for Instructors

The first source of trouble is the student spending too much time looking inside the cockpit. Watch what the student is looking at, out of the corner of your eye (looking straight at the student may be unnerving for the student....

I made it a habit to get students to look outside by covering up favorite instruments such as airspeed, or by putting my arm across the top of the instrument panel to make sure they don't look inside; the results are always better autorotations.

Another Teaching Point for Instructors

Obviously, an instructor experienced in autorotations is going to be more calm and cool than one who is not experienced. It is important that the instructor show calmness during this time in order to instill confidence in the student. Talking in a professional manner while performing the maneuver will go a long way to broadcasting that sense of calm.

Insurance (and other) Policies About Autorotations

Most insurance policies also have strong statements about autorotations.

It's also a good idea to remember this when you're on your own after you have your license. The policy of most helicopter flight schools is that they do NOT let students perform any autorotation exercise when solo for good reasons. If you're a student, it's best to observe this.

Many organizations do not permit touchdown autorotations except in regular check rides with an instructor. Another damper on this important skill is insurance coverage - most insurance companies do not like the added risk involved in practicing autorotations, mostly due to a bad history of accidents during this part of training. However, these problems do not reduce the requirement for training.

The end result is that many important judgment skills are not exercised regularly and pilots can become under-confident in their ability to safely handle an engine failure. How this problem can be solved will be discussed in Chapter 19.

The aim of autorotations is to arrive safely on the ground. Safely on the ground has one over-riding criteria - you can walk/hobble/crawl away from the helicopter. Even if the helicopter is not flyable; it's still a safe enough arrival.

Type of Helicopter Used in This Book

I've tried to make the instructions and examples in this book as neutral as possible about direction of rotation of the main rotor. For those times when I've had to use a specific direction, I've used the "North American" concept of the main rotor rotating counter-clockwise when viewed from above (or if you prefer- the blade at the front of the helicopter turning from right to left when viewed from the pilot's seat...).

Also, I refer to the slip ball throughout. If your helicopter only has is a slip string, use that instead. Obviously.

Simulators

For my money, the ideal place to start learning any aspect of helicopter flying is a synthetic training device (we'll call them all simulators for simplicity). Engine failure procedures is just one of those aspects that can be taught quite effectively in such devices. Chapter 19 deals with how to teach the fundamentals of autorotations in flight training devices / Flight Navigation Procedures Trainers / Simulators, as well as some of their limitations as trainers.

An Assumption or Two

I'm going to assume that as a helicopter pilot, there is no requirement to teach someone how to judge wind from naturally occurring sources, just as there should be no need to ensure that the wind direction is known at all times during the flight.

The wind is assumed to be constant throughout any maneuver.

It may be unnecessary to emphasize here that the instructor will demonstrate any of the maneuvers that follow first, with the student following through lightly on the controls, but I have to say it….

Summary of Chapter 1

This chapter has covered a lot of minor, but important points about the philosophy of teaching landings without the engine. Now it's time to get into the nitty-gritty details.

Chapter 2   Pre-Flight, Hover and Low Speed Training

Pre-Flight Briefings

Pre-flight briefings about engine failures and autorotations should follow the same format and layout as other pre-flight briefings - who is going to do what to whom when and where, what words are going to be used, and what safety items are essential.

The headings in the following chapters are a good starting point for items that need to be laid out to the student prior to getting in the helicopter for any exercise that has to do with engine failures.

New Terms / Limitations

Some new terms that will be used in this part of training might need to be emphasized. Some that have caused confusion are "go-around", "power recovery", and "overshoot". Make it clear which terms are going to be used, and then only use those words.

New limitations to be considered include:

- Rotor RPM limits for power off flight (Low rotor RPM and high rotor RPM)
- Low Rotor Warning lights and (and possibly) Low rotor warning audio
- V_{NE} in autorotation (we'll use $V_{NE\ Auto}$)
- $V_{Max\ Range}$ in autorotation (we'll use $V_{Max\ Range\ Auto}$)

Priorities

On the ground is also the best place to emphasize the priorities for helicopter flying that become even more important if the engine isn't available - namely, rotor RPM and then airspeed. Without rotor RPM no control is possible - it must be maintained at all costs.

Engine Deceleration Checks

It's worth checking that the throttle and fuel control on your helicopter is going to respond correctly when the engine is put to idle. More than one practice engine failure has turned into a real one because the instructor got a bit over-zealous with rolling the throttle to idle, and the engine failed[1]. It's worth doing this check with a bit of power on the engine, to simulate what will happen later in the training.

Rotor RPM Decay Rates

Once the engine is started and up to flight RPM, it's a good place to start showing how different power / collective settings will affect the rate of decay of rotor RPM once the engine stops. Start with the rotor at flat pitch and roll the throttle briskly and positively to idle[2] and note the rate of decay. Next, raise the collective slightly (about 25% of the way to

---

[1] Including me, I"m sad to say.

[2] Don't get too brisk because you might roll the throttle completely to off, which has happened to a few of us...

the hover in-ground-effect setting) and repeat the throttle reduction, lowering the collective immediately. Repeat but this time, don't lower the collective immediately and show how the rotor RPM will decay more quickly. This can be done several times with a different delay in lowering the collective, and with increasing power up to just below the light-on-the-skids point. There is no point in increasing beyond this point as there is a good chance the helicopter will start to yaw, especially when the throttle is reduced rapidly.

Best leave further examples to be shown when hovering.

It's worth emphasizing the change in noise at this point as well.

Attitudes - On the Ground and Hover

The first teaching point in the cockpit is to show two pitch attitudes - the "we're sitting on the ground" and the "hover" pitch attitudes, (if they haven't already been covered). It is worthwhile to point out where the horizon crosses the windshield center pillar when the helicopter is on the ground and again when hovering. This can be a location relative to a convenient rivet, or the standby compass, but should be something definite. This is one of the most important cues that tell the pilot the pitch attitude to expect on touchdown and in the hover (where the thrust vector is vertical to the ground and there is no motion either fore/aft or left/right). These two cues will be used at the end of the flare, and it is worth mentioning why it should be noted now.

It helps to have the student physically point this out by reaching out to the front and "marking" which screw/nut/other feature defines the two horizon-crossing points.

A variation is to have the student look straight ahead and note out of the corner of his/her eyes where the horizon is and how it appears to move up and down. At the end of the descent / start of the flare (covered later), this cue can be quite useful.

Figure 2-1 Hover Attitude.

Counting Down to Touchdown

This maneuver calibrates the seat--of--the--pants senses (proprioceptive cues for you techies) with where the ground is, or perhaps, where the skids and ground "interface".

9

While it should be second nature to all pilots (but especially to helicopter pilots) to be counting down to touchdown to know when the wheels or skids are going to contact the earth, this is as good a time as any to make sure it's being done. Starting from the hover, have the student count down from 3, so that the skids / wheels touch down at zero. This may take several touchdowns, but is very worthwhile, as it gives everyone a sense of the personal space of the helicopter.

This helps the student get over the "I know the ground is down there somewhere but I'm not sure where, and I know I don't want to hit it too hard" syndrome.

This is also a good warm-up for the instructor, especially if in a helicopter they are unfamiliar with. Moving to a helicopter with high skid gear when you're used to low skid gear could be a bit embarrassing....

No Robotic, Mechanical Control Movements (Please!)

One of the things to be emphasized is that every situation for landing following an engine failure is different, and that no one set of control responses is going to always be correct. No instructor should ever be demanding that the student must always add 2.5 inches of non-power pedal (as just one example). I remember one military organization that insisted that in a hovering autorotation it was essential to always push forward on the cyclic, when no such movement was needed.

Rather than have the student repeat words like "Down, Right, Aft", which starts the student thinking (and acting) mechanically, use terms like "Stop the drift", "Stop the yaw" and so on. If you become "mechanical" then it becomes difficult to change -- for example to helicopters with the main rotor direction opposite to what you trained in.

Hover Engine Failures

It's incorrect to call a landing from an engine failure in the hover an "autorotation", as the air doesn't have a chance to keep the rotors turning - so, to be correct, they'll be called "hover engine failures".

This section is going to dissect this important building block in great detail, as it is a foundation for all other parts of landing without power.

First- Stop the Drift

What happens to a hovering helicopter when the engine fails? First thing that <u>appears</u> to happen is the helicopter yaws, as the requirement to counteract the power to the main rotor has disappeared. How much it yaws depends on the power being used, relative wind speed and direction, and so on. You'd think we'd worry about the yaw first, but there's a good reason why we don't. Drift is more important, even if you don't notice it right away.

To understand the reason for the drift, it's necessary to understand how the main rotor thrust is tilted to overcome the thrust of the tail rotor. That's a basic feature of all helicopters, and you'll have to look in other books[1] to find that.

As both main and tail rotors are slowing down rapidly, the effectiveness of the tail rotor is going to reduce. This has no real effect on the ability of the tail rotor to stop the yaw, (the

---

[1] Like ""Cyclic and Collective"

main rotor reaction is likewise reducing), but it has a marked effect on the translating tendency. Typically the helicopter drifts to the side slightly immediately following the engine failure[1]. The aim of the pilot is to stop the lateral drift. There may be some longitudinal drift (depending on the wind), which is OK as long as it is a drift forward.

It's important to stop the lateral drift, as it can tip you over if you hit the ground moving sideways in a skid equipped helicopter. That's the reason why the first priority in an engine failure in the hover is to "STOP THE DRIFT".

Fortunately, it only takes a small amount of lateral cyclic to stop the drift and it's almost instinctive to make this correction once you have been alerted to it. It should nevertheless be pointed out and emphasized as the first priority.

And since there may be times when the helicopter drifts in a different direction than taught in training due to winds (or a rotor system turning in the other direction), it is important that the concept be "Stop the Drift" rather than "add left (or right) cyclic".

Stop the Yaw

The amount of yaw is of less importance than the visualization of "STOP THE YAW". It's OK to touch down with some yawing motion (compared to drifting sideways). And don't try to return to the original heading when the engine failed, as if the engine failure is real, there will be some considerable yaw before the failure is recognized. As one of my friends noted, the amount of pedal movement needed is huge.

Having stopped the drift and yaw, the next aim is to touch down at an acceptable rate (using the collective lever as necessary).

The published not-so-practical test standards in one country concentrate on the pilot maintaining heading precisely in a practice engine failure.

That would be nice if engines provided a warning of impending failure, but they don't. In a real engine failure, there will be a change in heading, and your task is merely to stop the yaw, not return to the original heading. As one experienced instructor said of this (im)practical test standard - all it shows is that the student can do three things at once.

Cushion the Touchdown

The last part of the maneuver is to control the rate of descent using the energy stored in the rotor RPM. The best words to describe this part of the maneuver are "Cushion the touchdown". You will be landing, gravity ensures that. The question is "how smoothly do I want to touchdown?".

Judging Collective Lever Application

Some students have difficulty judging collective lever application.

Like many other things, they need experience of how to use the rotational energy, when to use it, and how rapidly to use it. Before we get into the nuts and bolts of how to use rotational energy, we need to discuss what it is.

---

[1] You'll note that there is no mention of whether the drift is to the left or to the right. We're trying to teach principles, not robotic mechanical control applications.

Rotational energy is the energy stored in the rotor due to its whirling around. Like kinetic energy, it depends on a square function: the energy of a rotor turning at 100 RPM is four times the energy of a rotor spinning at 50 RPM. Since we can extract this energy and turn it into lift, and the amount of lift produced will also vary as the square of the rotor RPM; it doesn't take much mathematics to show the lift available at the high end of rotor RPM is much greater than at slower RPM. If you're trying to stop the helicopter descending, it will take a lot more collective lever movement to generate the lift if you start at a low rotor RPM than at high rotor RPM.

All this complicated explanation is to show it's difficult to determine how to use this energy!

Falling from 6 Inches Above the Ground

Here's a method to help show this. Start with the helicopter at a very low height (less than one foot above the ground) and maintain it there after the engine is "failed" using increasing collective lever, until there is no more lift. At some point, as the collective lever is being increased and angle of attack is increasing towards the stall, the rotor will stop producing lift. When the rotor stops "flying", the helicopter plummets the short distance without harm.

This teaches a lot about controlling collective lever inputs! A couple of attempts at this will show the student how to apply the correct amount of collective.

Once the helicopter is close to the ground, the collective should only move up. If you've pulled too hard, don't push down -- wait.

Running Landings

...With Power

The next part of the warm up is to conduct running landings under power at slow forward speed. Again, the aim is to judge exactly when the undercarriage is going to touchdown. The reason is the same as the vertical landing. As part of the discipline needed for the next exercises, do not lower the collective until the forward motion has stopped. The transfer of weight from the rotor to the skids should only be done as gradually as possible to stop the helicopter from "digging in", particularly if you end up landing on a soft surface in a real situation (see Chapter 11 about why your training autorotations should be only be done on smooth paved surfaces).

Why Smoother Touchdowns?

You'll probably notice that the touchdowns from a hover taxi are slightly more controlled than from a stationary hover. The reason for this is that we have eyes in the front of our heads and get better depth perception from forward motion than moving in a purely vertical sense.

Hover Taxi Engine Failures

Hover-taxi engine failures are the logical next step and are so similar to the hover engine failures that there is no need to repeat the steps.

The assumption in this phase is that the height in the hover-taxi is below the Height Velocity (H-V) curve low hover height. There should be no attempt to decelerate between engine failure and ground contact, and there should be no (or very little) attempt to slow using the cyclic after ground contact.

Also, just like the running landings with power, do not lower the collective until the forward motion has stopped.

Engine failures in a hover-taxi are not much different than in the hover. The aim is to stop the lateral drift, stop the yaw and then cushion the touchdown.

For this reason, it's a good idea to maintain the skids aligned with the direction of travel when hover taxiing -- it's one less thing to worry about.

Hover Height and the H-V Curve

Chapter 14 will deal with the low hover point of the H-V curve in much more detail, however at this stage, the student should be encouraged to hover no higher than 6 to 8 feet above the ground (unless there is a very, very good reason for it).

While it is possible to conduct training engine failures in the hover above this height, a real engine failure above the low hover height on the H-V curve is very often different than those seen in most training.

Remember the earlier comments about students and training scenarios -- a training flight carries the implication that engine failures will be part of the trip, and "real" flights don't.

Slightly Higher Hover Engine Failures

Sadly, not all hovering takes place at 3-5' above the ground. So how do you handle an engine failure if you're 15' above the ground (despite advice you'll get later in The-H-V-Low Hover Point) and the engine quits?

This calls for a different technique than the low hover method. A lot of things are going to happen -- simultaneously. You won't be able to let the helicopter descend slowly while raising the collective or even holding the collective lever in the same place.

It will be necessary to provide a sharp downward movement of the collective lever to persuade the helicopter to start down. How much down collective lever and how rapidly it needs to be added can only come from experience.

The aim is to use the rotor's precious rotational energy and thus, lift, in the wisest possible way. If you're going to practice this, it must be with an instructor who cherishes the energy in the rotor dearly.

As the helicopter nears the ground, a rapid up-application of collective lever will be needed to stop the rate of descent. How much and how rapid will depend on the situation. Just remember any energy remaining in the rotor after touchdown doesn't do you any good if the touchdown is going to be threatening to your physical body.

The problem with this method is that real engine failures don't provide enough warning to allow the downward check of the collective to be used unless you're hovering at some considerable height above the ground.

All real engine failures that I know of (from my own experience and from talking to others who had them) result in the pilot only having enough time to raise the collective, and as will be seen in the chapter on the low hover part of the H-V curve, this is the technique that should be taught for hovering engine failures close to the ground.

After flying a number of H-V demonstrations at the low hover point, I cringe whenever I see single engine helicopters hovering needlessly above that height. Especially students…

Note that this is something you don't - repeat don't - want to practice as there is very little margin for error. I know….

Self-Initiated Engine Failures in the Hover

Typically, when asked to initiate an engine failure in the hover, many students spend quite a while "getting ready", and are slightly afraid to close the throttle themselves. It may help if the instructor closes the throttle without much warning to show the student they really were ready for the failure at any time.

While the student is getting ready to roll the throttle off, roll it off without warning. Tell the student that they were ready all along but just didn't realize it.

Practice until Instinctive

Practice the engine failure in the hover and hover-taxi until the actions become instinctive. The student should be able to discuss the sequence of events, and self criticize his performance.

Quick Stops

Quick stops are practiced for several reasons. First, they duplicate most of the flare part of the autorotation quite well. Second, they are an excellent coordination exercise for beginning pilots. See Figure 2-2 for the sequence of events in the quick stop.

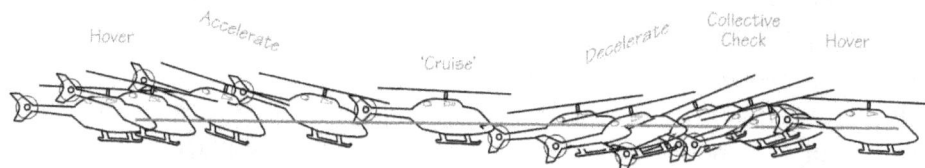

Figure 2-2 Sequence of Attitudes from Hover to Hover.

To fully understand the reason for practicing the quick stop, consider the main purpose of the flare in the autorotation: to stop the rate of descent. A secondary reason is to decrease ground speed, but the main purpose is to stop the rate of descent. A beneficial side effect is the increase in rotor RPM from more airflow through the disk.

The first quick stop should use a slow airspeed and gentle flare.

This is where an intuitive knowledge of the "personal space" the helicopter occupies is needed. Even though the tail rotor is a long way behind you, you should know where it is all the time, and know how much clearance it has from the ground.

Start with a "cruise" segment of about 20 knots[1] at 20' AGL. When the deceleration is controlled and coordinated, then use 30 knots. And so on.

On subsequent quick stops increase the airspeed as well as rapidity and nose up attitude of the flare until the collective lever is nearly full down in the flare. In all of them, practice maintaining a constant height above ground during a quick stop, and of course, control the

---

[1] The airspeed indicator won't be of much use, learn to judge 20 knots by the rate of the ground going by

heading. Concentrate on maintaining altitude with the cyclic stick until the cyclic stick is ineffective in holding the helicopter at a constant height, that is, bringing the nose up won't keep the helicopter at the desired height above the ground.

For ex-military types, the quick stop is different than what you may have been taught. This is not a mechanical maneuver where collective is added first.

Watch out to ensure the tail stinger doesn't hit the ground -- but you should know the "personal space" the helicopter occupies instinctively by now.

The next control input is one of the secrets for a successful autorotation: the helicopter must make the change from a nose up, decelerating attitude to a hovering pitch attitude for touchdown. How this is accomplished has a great bearing on the quality of the touchdown. In nearly all helicopters the cyclic stick is not the best control to level the helicopter at the end of a flare!

Keep the Skids Straight

One of the more difficult things for student pilots to master in this busy phase of flight is the requirement to keep the skids aligned with the direction of flight. A useful discipline that will pay off later.

Downwind! Not

Don't do quickstops with an aim to stop facing downwind for this training exercise. While never a good idea[1], it's a very bad idea for this series of learning points.

Collective Check - Why It Works

Assuming you had enough airspeed to flare, and the helicopter is at the end of the flare, with very little forward speed relative to the ground, and is in a nose high attitude -- now it's time to get the fuselage level and land. As previously mentioned, the best way to go from the end of the flare to the touchdown is via the collective lever check. The only helicopters I have seen where the collective lever check did not work were tandem rotors, but for nearly every single-rotor helicopter, it is the best method to get the helicopter level.

Why is the collective the answer for leveling the helicopter? Isn't the cyclic normally used for that type of motion? A large amount of forward cyclic stick is needed to generate the pitching moment needed to level the helicopter in autorotation, compared to that needed in "powered" flight, as the thrust vector is very small.

At the end of the flare, the collective lever should be nearly fully down. The helicopter is in a nose up attitude, and the ground speed is low. A small but positive check up in collective lever increases the main rotor blade pitch, and hence the thrust, at the expense of some rotor RPM. The effect of the increase in thrust with the nose up attitude creates a moment (a technical term for force acting at a distance) about the center of gravity. Since the CG is normally forward of the rotor mast, the natural effect is for the thrust vector and the weight to try to align themselves, bringing the nose to a nearly level attitude, without any movement of the cyclic stick. This leveling of the nose is self-correcting; normally, it stops the helicopter at the disk level attitude. As the disk approaches the level position, the couple becomes

---

[1] In some countries a quickstop started while traveling downwind but ending up facing into wind is a required maneuver for licenses

weaker, and when the nose is nearly level, the couple has diminished. This is shown in Figure 2-3.

The collective lever check can also be considered as using the rotor disk as a giant air brake.

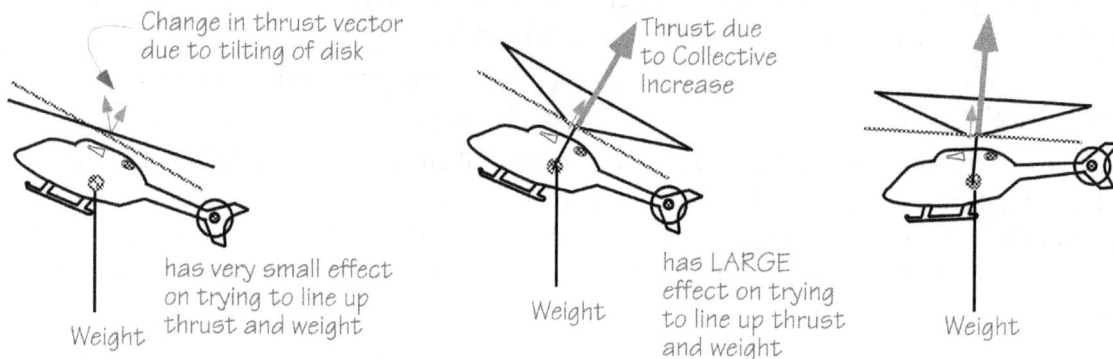

Figure 2-3 The-Collective-Check.

Quick-stop the Wrong Way

Try doing a quickstop the "wrong" way to show this important point.

Use the following control sequence: at the end of the flare, as the ground speed comes to zero and the helicopter starts to sink, level first with the cyclic stick, and then add collective lever to stop the descent. The results will be less than satisfactory, particularly compared to the "proper" sequence (collective lever first as a check, and only if this check does not bring the helicopter to the level position, make a small adjustment with the cyclic stick). Even non-pilots recognize this as an ineffective way to get to the hover attitude.

The reasons for not using the cyclic stick to level the helicopter are many:

- The cyclic stick is not very effective, as it is only changing the direction of the very small thrust vector. It takes quite a lot of cyclic stick to level, if only the cyclic stick is used.
- Forward cyclic stick introduces forward speed again. One of the purposes of the flare is to reduce ground speed, and using forward cyclic stick undoes this work.
- Forward cyclic stick introduces a nose down pitch rate. This pitch rate continues past the "hover" attitude, and must be stopped prior to touchdown, particularly on soft ground, in order to prevent the nose from digging in. The only way to stop this pitch rate is with aft cyclic stick, leading to a lot of cyclic stick activity at an already busy time.
- Forward cyclic stick bleeds rotor RPM unnecessarily. When the helicopter is flared and the cyclic stick pulled aft, the rotor RPM increases - the opposite happens when the cyclic stick is pushed forward - the rotor RPM decreases.
- Finally; the cyclic movement is seldom (if ever) needed.

Using the collective lever to put the helicopter in a level attitude has none of the disadvantages of using the cyclic.

A note of caution: remember the earlier caution about stopping a quickstop with a tailwind, and especially don't try too rapid a flare downwind, as it is possible to get into vortex ring state in this condition, especially if a rapid collective lever check is used at the end. Since the quick stop maneuver is also a key to many other things, again shows the sequence of events from the start of the quick stop to the hover.

Coupling of Forces in Leveling a Helicopter

I once had occasion to convince an unbelieving student that the collective lever would do all that was needed at the end of the flare (he was a high time tail wheel fixed wing crop sprayer who was pulling back on the cyclic after the end of the flare). At the end of the flare, I took my hand off the cyclic stick and leveled the helicopter and completed the touchdown without using the cyclic stick. I do not recommend this as a normal method, (the conditions were ideal) but it certainly got the message across!

Why does this work? In most helicopters, the CG is ahead of the rotor mast, and it's natural for the CG to want to line up with the lift vector from the rotor (even though it is small).

If the CG is very close to the rotor mast, then the collective lever check may also not be as effective. If the CG is a long way aft of the rotor mast (not common in single rotor helicopters), then the check could have the opposite effect to that desired.

Your task is to get the helicopter to the hover pitch attitude using the flight controls. If you are well practiced, then a combination of collective lever check and cyclic stick will obviously be the best solution, but remember raising the collective lever will increase the size of the thrust vector the cyclic stick is tilting.

See Figure 2-3 for the way the coupling of forces works when a "collective lever check" is used.

Well, so far, we haven't even gone more than 10 or 15 feet above the ground, and we've learned a lot. Next, we go higher.

Summary of Chapter 2

This chapter has covered how to handle an engine failure in the hover and hover-taxi. Students should be prepared to go to the next level when they have mastered these exercises.

Questions

1. What are the three things that need to be done following an engine failure in the hover or hover taxi?
2. Should you try to slow the helicopter's groundspeed following an engine failure in the hover-taxi?
3. Why should you keep the skids aligned with the direction of travel when hover-taxiing?
4. Is it prudent to hover above the height shown as the low hover point on the H-V diagram?

Chapter 3   Upper Air Work

While the landing is the icing on the cake following an engine failure, a controlled descent is a necessary precursor to getting to the flare and landing. Flight without an engine is a basic procedure that every helicopter pilot should know and understand intimately.

Autorotation is the condition of flight where the rotor is driven by aerodynamic forces, as opposed to power from the engine. The air coming up from beneath (or the air through which the helicopter is descending, depending on your point of view) provides the forces to make the rotor go around. Ground school theory outlines how this happens.

Aside from the beginning of the emergency (the surprise of having lost power from the engine) and the end (the need to arrive safely on the ground), autorotations are not difficult.

Unfortunately, during most training, the majority of the flying time is at low height above the ground. Any practice in landings without the engine is from this low height, with a good deal of pre-built stress, and little time spent in a stable descent. The event is over almost before it's started --- from 500' above ground level (AGL), it's often less than 30 seconds (of panic by the student) before the helicopter is on the ground.

- Some fundamentals need to be emphasized.
- The helicopter can still fly despite the fact that the engine is not turning the rotors.
- The helicopter remains fully maneuverable as long as the rotor RPM is within limits.

Using this long-ish descent will significantly increase the amount of time spent in an autorotational descent, plus allow very specific points to be covered. These specific points will have a large impact in later autorotations.

Prior to the Exercise

Prior to conducting this exercise, review (memorize) the manufacturer's recommended airspeeds and limitations for minimum and maximum power off rotor RPM.

Airspeeds to be used in this exercise start with recommended airspeed for autorotation and use the minimum rate of descent airspeed, maximum range airspeed in autorotation down to zero airspeed.

Entry to autorotation will be covered later - the main part of this exercise is the steady state portion of the descent. Other parts of the autorotation will come later and will be repeated often, but this part concentrates only on the descent.

The instructor should ensure that the student knows the exact words to be used for simulating the engine failure, and other important words for controlling attitude and rotor RPM.

If you're the instructor, it's of little use to say to the student "Airspeed, Airspeed" -- the student doesn't know if it's too high or too low, merely that you're not happy with it. Try "Increase Airspeed" or "decrease rotor RPM" as positive things to do.

On the other hand if you are trying to make the more advanced student sort out the problem themselves, you might be better off to just say "Airspeed".

Checks

Select an altitude that will provide sufficient height so that things will not be rushed, and that will give plenty of time to experience the flying characteristics of the helicopter in autorotation. For this exercise, this should be several thousand feet above the ground, probably higher than used in other exercises. Enjoy the view! The time spent in climbing can be used to improve basic skills such as maintaining airspeed.

Establish the helicopter in straight and level flight at best rate of climb / minimum rate of descent / minimum power airspeed (which will be shown as the shorthand of V_Y for the rest of the book. During this exercise, the student might be spending too much time looking inside at instruments, rather than looking outside, and both of you can get focused on something other than looking out for other traffic. It's prudent to make sure you're not going to be a threat to anyone else in the vicinity (or that they won't be a threat to you!).

Good airmanship requires that prior to carrying out an autorotation descent, the necessary "pre-entry" checks be completed. These include:

Normal Pre-landing check

The normal pre-landing check should be done now, as well as something called the HASEL check.

HASEL check

H - Height sufficient for exercise;   (sufficient altitude for the landing spot you want)

A - Area good for landing   (if required);

S - Security (or perhaps Safety)   (straps, no loose articles such as kneeboards or maps that could drift around and distract you at an important time.); this should be the normal state of affairs anyway.

E - Engine -   Temperatures and Pressures; and

L - Lookout - Look around, especially below in the direction of any expected turns. Check the wind direction on the ground if you can. Helicopters or airplanes on the ground should also be considered.

There should be a verbal indication from the instructor of what is about to happen. These should be the same words as used in the pre-flight briefing.

Leave Yourself an Out

It's always worthwhile to leave yourself an "out" in the event that something untoward happens. What I mean by this, is that if things do go wrong and you need to complete a real engine off landing because the engine didn't respond in the recovery (or quit when it was put to idle[1]), it's useful to make sure that there is somewhere suitable to land.

I don't mean you have to be over an airfield with paved runways, but somewhere at least that's hospitable to landing helicopters.

Once the pre-entry (HASEL) checks have been completed, the helicopter can be set up for entry to autorotation.

---

[1] …happened to me!

Gentle Entry To Autorotation

This exercise is to show handling in a descent in autorotation and not the actions immediately following an engine failure. The transition from powered flight to autorotation in this exercise will be is slow and deliberate and very different than that used in future exercises.

The wording for this exercise might be "Practice Autorotation" - but whatever wording is used, should have been briefed beforehand.

In this case, starting from V_Y, the collective will be lowered before the engine power is reduced to idle and put the slip ball in the middle. In all probability, some aft cyclic and non-power pedal will be needed.

In most helicopters, the collective can be lowered completely, but at this higher-than-normal altitude, with less dense air, care must be taken to ensure the rotor RPM doesn't exceed the power off limitations.

A slight amount of collective pitch may be needed to maintain the rotor RPM within limits. Have the student note the collective position to maintain the rotor RPM in the middle of the power-off range.

Maintain the airspeed at V_Y. Slight changes in cyclic and pedal may be required to maintain the airspeed and slip ball centered.

Once the rotor RPM is controlled, put the engine to idle, and briefly note the engine indications such as the engine idle RPM. The rotor RPM and engine RPM needles should be split.

The student will already be familiar with attitude flying, and the instructor should note the pitch attitude relative to the horizon.

If the pitch attitude remains constant, the airspeed doesn't change.

It's worthwhile to have the student reach out and point to where the horizon cuts across the windshield center support.

For Those with Governors on their Piston Engine Helicopters

Those fortunate enough to have a governor to help maintain rotor RPM in their piston engine helicopters should know how it's going to react with these power changes- both entry to autorotation and recovery from it.

Change Rotor RPM

Once stabilized on airspeed, makes some small changes to the collective position (if you already have the collective bottomed, you'll obviously only be able to raise it...) and note the change in rotor RPM.

Slowly raise the collective to reduce the rotor RPM until the low rotor RPM warning light and horn are activated (if fitted) and then further (if possible) to the minimum power-off rotor RPM value, and then lower the collective again to show how the rotor RPM returns to the original value. If you have to hold the collective pitch away from the bottom stop, you can lower it until you get to the maximum power-off rotor RPM. Notice the difference in noise from the rotor at the different rotor RPMs.

Be (relatively) slow and smooth making collective changes. Maintain airspeed and keep the slip ball centered.

Stress how minor changes in collective position are used, and that the effects sometimes take a few seconds to become apparent.

Repeat with slightly more rapid movements of the collective.

Maintain Airspeed

The instructor may cover up the airspeed indicator now; it's for a reason. Look outside and note the attitude. Don't change it and the airspeed won't change. Maintain attitude = constant airspeed.

Listen for the noise of the rotor. Note the rate of descent, and if you're up to it, record the rate of descent when the airspeed is stable.

Change Airspeed in Autorotation

Once the rotor RPM has stabilized, increase airspeed to the recommended best range airspeed ($V_{Max Range Auto}$). Once stabilized at that airspeed with the ball centered, note the more nose-down attitude and how slightly more non-power pedal is needed. Note that it may take a moment for the airspeed to stabilize, however, if the new attitude is maintained the airspeed will once again remain constant. Repeat the changes in collective to see how the new airspeed has a slightly different effect on changing rotor RPM.

Set the collective to get "normal" rotor RPM again and increase airspeed to $V_{NE Auto}$. Note the more nose down attitude and the change in pedal necessary to keep the slip ball centered.

Try using a change in pitch attitude to return to the original airspeed (the "normal" airspeed for autorotation). The best way to do this is to cover the airspeed indicator and once the pitch attitude is stabilized, then uncover the airspeed indicator to show how close to the desired airspeed the helicopter is.

Low Airspeed / Zero Groundspeed Descent

It should be possible to determine the wind direction at altitude (either from a pre-flight check of the wind at altitude from the weather office, or if the conditions permit from cues such as cloud shadow movement across the ground).

Turn into the wind at altitude and reduce to zero (low or nearly zero) groundspeed (note groundspeed). In order to get zero groundspeed, it will be necessary to look out the side window, and slow the helicopter until progress across the ground is effectively stopped. We're not trying for zero airspeed - you can't measure it anyway. Note how the rotor RPM is not affected by this low airspeed. To recover to "normal" airspeed in autorotation, make a smooth and positive, but not overly aggressive forward cyclic movement to regain airspeed. If there is an attitude indicator in the helicopter, you can target about 10° nose down and note how long it takes for the airspeed to start to indicate and then to finally settle at "normal".

More Rapid Changes in Airspeed

Once the fundamentals of changing airspeed in autorotation are mastered, try a bit more rapid change. Note the rotor RPM reaction to a nose up (reducing airspeed) movement as opposed to a nose down (increasing airspeed). A nose-up pitch at any reasonable rate of pitch change will cause the rotor RPM to increase momentarily, while pushing the nose over will cause it to decrease momentarily. Also note that changing from low airspeed to higher airspeed takes a lot altitude and time to get sensible airspeed changes. File this useful fact for later!

Turns In Autorotation

Turns in autorotation are not much different than turns in powered flight, with one small change.

Just as the rotor RPM increases while the nose of the helicopter is being raised, so will the rotor RPM increase when the helicopter is turned. Listen for the increase in noise, and note it on the RPM gauge.

As would have been shown in earlier flights, angle of bank is not the only thing required for a turn: the nose needs to be tracking across the horizon, which requires some aft cyclic. This is the same as pitching the nose up (as was done to reduce airspeed), and the effect on the rotor RPM is the same: it will increase slightly. If a steeper turn with more G loading is made, the rotor RPM will increase more, and a larger increase in collective may be needed to maintain the rotor RPM within power-off limits. Try some turns up to 60° of bank to show this.

If the rotor RPM increases too much, raise the collective slightly to keep it within limits.

When you roll out of the turn, it may be necessary to lower the collective to keep the rotor RPM from decaying. Turn in the other direction.

A few turns and this will become second nature, as it should be.

Ball (or String) Centered

Try a small change in slip ball position. To maintain a constant airspeed a slight amount of forward cyclic is needed to overcome the increased drag caused by the greater side area presented to the air. Note how the rate of descent increases at the same airspeed when the slip ball (or slip string) is not centered.

Flying with the ball out of center of the inclinometer causes unnecessary drag and is aerodynamically inefficient. It will adversely affect the glide distance.

On the other hand, if you're too high and too close to the only suitable landing site, this may be useful information.

Track Across the Ground

At least one helicopter type will show an interesting effect in autorotation: with the slip ball centered and the "wings" (rotor or airframe, if you wish) level, the helicopter will be moving laterally across the ground. This may not be evident at high altitude, but will definitely be noticeable when closer to cues. Some helicopters have more lateral drift than others.

Your task as the pilot is to make the track across the ground what you need it to be by adjusting heading (with the slip ball centered).

A Note about Rotor RPM and Altitude

The rotor RPM in autorotation changes depending on a great number of variables. One of these is density altitude. For this exercise, the density altitude will change a great deal compared to most of the rest of your training.

Initially, at high altitude, the rotor RPM with the collective fully lowered will be higher than it will be when closer to the ground.

It may be too high (outside the limitations for power-off rotor RPM) and it may be necessary to raise the collective slightly to keep the rotor RPM within limits. This is especially true on rotor systems with more than two rotor blades. It may not ever be possible to fully lower the collective.

Recovery to Powered Flight

This should be done well away from the ground for this exercise - the student is still trying to learn just about handling in autorotation and recovery from it. No sense confusing the situation with trying to end the descent with a hovering power recovery. I'd recommend no lower than 500' AGL for this return to powered flight.

To overshoot[1] from the autorotation descent, smoothly increase the throttle to rejoin the needles within the correct operating range. The techniques will be different for piston and turbine-powered helicopters that are covered later in Turbine Engine Power Recovery.

Be prepared to raise the collective to prevent the rotor RPM from overspeeding.

Once the needles are matched, raise the collective to set climb power and at the same time adjust the cyclic to achieve the best rate of climb.

Anticipate the requirement for power pedal to maintain coordinated flight throughout the transition from autorotation descent to powered flight.

Once established in a stabilized climb, perform a post-takeoff check.

Ensure temperatures and pressures are in the normal operating range.

This is important because the engine has been operating at idle for a long while -- piston engines are particularly cranky about large power and temperature changes and shock cooling.

The opportunity to practice entries, descents, and overshoots will come later.

This whole sequence can be repeated a couple of times until the student demonstrates proficiency and understanding of how changes in collective position, pitch attitude, airspeed, angle of bank and pedal position are inter-related.

Summary of Chapter 3

The exercise in this chapter will provide the student with significantly more flight time in autorotation than would be found if the only exposure were to autorotations from 500' above ground. With this exercise, the student will feel much more comfortable in knowing how to control the helicopter in autorotation.

Questions

5. What will happen to the rotor RPM when the helicopter is turned in forward flight in autorotation?
6. What happens to the rotor RPM in a steep turn in forward flight in autorotation?
7. Will the rotor continue to turn at zero airspeed in autorotation?
8. What is the effect on the rotor RPM of lowering the nose to increase airspeed?
9. What is the effect on the rotor RPM when the nose is raised to decrease airspeed?
10. If the pitch attitude is kept constant in autorotation what happens to the airspeed?

---

[1] or go-around, depending on your terminology. Make sure the crew knows which words will be used.

Chapter 4   The Entry to Autorotation

The entry is taught after the steady descent portion, as it is less alarming when you know what is going to be happening next. Trying to teach the entry and then the descent at the same time is slightly counter-productive.

The time between the engine failing (or simulating the engine failure) and when the collective lever must be lowered is very brief. The student must learn the symptoms of an engine failure and the correct reactions to be able to respond instinctively.

When Practicing....

It's useful for students to know whether the sudden change in events that happens is a training exercise, or a real engine failure. At some point either immediately before, or during the simulation, the instructor should state: "Practice engine failure" as this clearly describes the situation.

Symptoms of Engine Failures

The symptoms of an engine failure can be perversely vexing. They may not be immediately noticeable, or they can be alarming. Know them for your helicopter type. One way to learn them is to read accident reports.

Proprioceptive (seat of the pants) Cues

Nearly all engine failures effect the balance of the forces, especially between the main and tail rotor, so it should come as no surprise that one of the first and largest cues to the pilot will be a yawing motion, even in forward flight.

A second cue is that the helicopter may sink slightly as power is lost and the main rotor begins to slow. This is the second large cue that something is amiss.

Noise

In some helicopters with noisy piston engines, the sudden lack of noise from the engine is an extremely strong cue. In other helicopters with well-insulated turbine engines, it may be very difficult to hear any change in engine noise.

Immediate Actions

When you have practiced entry to autorotation more than a couple of times, the following will seem as natural as walking. At first it will all seem a bit rushed, and will need to be practiced until it becomes second nature. Hence the necessity to slowly go through the steps.

First - Move the Cyclic Aft

It is difficult to split this step from the next one -- the cyclic and collective can (and should) always be moved together when entering autorotation. How much the cyclic needs to be moved aft will depend on the situation, but it should be moved aft to start the air flowing up through the rotor.

The reason for first moving the cyclic aft is that rotor RPM is much more important than airspeed, and that reducing collective on it's own will tilt the disk forward. Any forward tilt of the rotor disk will reduce the upward airflow through the rotor disk. Upward airflow through the rotor is essential to maintaining rotor RPM.

In some cases, such as in the cruise, moving the cyclic aft may be accomplished a long time before it is necessary to lower the collective.

But whatever it is, moving the cyclic aft should be the first and instinctive movement. There is only one time when it's not necessary -- when you're in a hover, especially close to the ground.

And if you push forward on the cyclic immediately after the engine fails, you'll just be going faster towards the ground.

At the Same Time Reduce Collective

The collective needs to be lowered quickly and (in most training helicopters) completely. This will reduce the drag on the main rotor blades and permit the aerodynamics to take over the task of providing the necessary forces to keep the rotor turning.

Depending on the helicopter, after the initial lowering of the collective, it may be necessary to raise it slightly to prevent the rotor from overspeeding, but this depends on the helicopter model.

Add the Necessary Amount of Pedal

Since the engine is no longer driving the rotor, the necessity to counter-act its torque has also been removed. So it's necessary to use the pedals accordingly to try to get the slip ball into the middle (for those who only have slip strings, - put the slip string in the middle). But for the moment, don't be too picky on getting the ball exactly centered. That can be done in a few moments when things have settled down slightly. You will quickly become attuned to how much non -power pedal is needed.

Airspeed above V_Y

The second thing that should be done, if entering from an airspeed above the best rate of climb airspeed (V_Y) or the recommended airspeed for autorotation, is to reduce the airspeed towards the optimum desired for the descent. This should be started at the same time as the collective is lowered. I know of no helicopter where it is necessary to add forward cyclic following an engine failure at an airspeed above V_Y.

Airspeed Below V_Y

If you're unfortunate enough to have a low airspeed (but plenty of height) when the engine fails, then, once the rotor RPM is in the green and stabilized, (and only then) should you put the nose down to gain some airspeed. But not initially -- wait until the rotor RPM has at least settled near the desired range. Pushing the cyclic forward with the rotor RPM low will only cause it to go lower. Airspeed is not as important as rotor RPM.

Rate of Decay of Rotor RPM

The rate at which the rotor RPM will decay following an engine failure depends on the power demanded of the rotor and the flight condition.

For example, the rotor RPM will decay very quickly if in a high power vertical climb out of a confined area, but will decay very slowly if in a descent at 60 Knots at (needless to say) low power.

For this reason, it's worthwhile to spend some time showing this to a student. The following build-up is a good way to do this.

Start in level flight at the airspeed that requires the lowest power (typically 50 to 60 knots in most light training helicopters). The student should already have been briefed about how this exercise will progress. Tell the student that initially you want the collective lever to be lowered immediately when the throttle is closed. Note how low the rotor RPM drops. Recover to powered flight in the same airspeed and power setting.

Next, when simulating the engine failure, delay lowering the collective for a short (very short) time (but do please move the cyclic aft).

Note how quickly the rotor RPM decays. Smartly (quickly) lower the collective, and bring the engine back on line and get back to level flight. Repeat the exercise, but this time add a slight amount of power so that the helicopter is climbing. Repeat with slightly more power so that the rate of climb of greater than before.

Review of What to Do

In a real engine failure, you'll be presented with a lot of information.

Not all of it will make sense, and there will also be a huge "this can't be happening (to me, today)" sensation.

The things that aren't "right" --.

- Yaw without any input,
- sinking feeling,
- engine noise (or lack thereof),
- rotor slowing down noise,
- horns may be sounding and
- lights flashing.

You have a bunch of things that aren't right, and you need to manage them. Sort out your priorities -- maintaining rotor RPM is the highest.

That is best accomplished by aft cyclic, down collective, use the pedals to correct for yaw, look for place to land (more on that later).

Summary of Chapter 4

The entry to autorotation must be an instinctive maneuver and entered whenever the symptoms of an engine failure present themselves.

Questions

1. What are the main symptoms of an engine failure in the cruise?
2. What are the immediate reactions for an engine failure in the cruise?

3. When should the cyclic be pushed forward in an autorotation entry?

4. When should the cyclic not be pushed forward in an autorotation entry?

5. What are the symptoms of an engine failure in the hover?

6. Under which flight conditions should the cyclic be the first control to be moved following an engine failure (real or simulated)?

7. What are the priorities with respect to airspeed and rotor RPM following an engine failure?

Chapter 5   The Flare

Logically, we should talk about the descent and the variations on the descent at this point. However, since that has a lot to do with performance, we'll leave that for a while. If you wish to read it now, it's in Chapter 9. More about energy is in Chapter 6.

Between the descent part of the autorotation and the landing lies the flare, one of the most misunderstood parts of the autorotation.

This misunderstanding has a very long after-effect.

The main purpose of the flare is to STOP THE RATE OF DESCENT.

Nothing else.

It's not to slow the airspeed, or to build the rotor RPM, although those things can and probably will happen. The main purpose is to stop, (or at least attempt to control) the rate of descent. There are two other useful side effects - it does slow the forward speed to make the landing less dramatic, and it increases the rotor RPM.

The purpose that must be remembered is to stop the rate of descent.

Keep this foremost in mind and autorotations are much easier.

There is no textbook, by--the--numbers procedure for how much to flare, where to flare, or when to stop the flare. This is a visual maneuver (but see Chapter 20 on night and instrument autorotations), and requires some judgment. Judgment implies experience, and hopefully a variety of experience.

Some flight manuals (most military ones, it seems) have precise wording the pilot is expected to be able to remember at an infrequent moment of extreme stress; for example:.

> *At 100 feet above ground, flare the helicopter to 10-12° nose up, using approximately 2" of aft cyclic stick. At 25 feet above ground, level the skids and apply collective lever to control the touchdown.*

How absurd.

In an attempt to reduce everything to an absolute minimum robotic level of performance, the adaptability and judgment of the human pilot has been sacrificed.

The Correct Attitude at the End of the Flare

There are at least two correct attitudes to be in at the end of the flare. One is the real-world correct attitude, and the other is the it-would-be-nice attitude. The main purpose of the flare is to stop the rate of descent, slowing the helicopter's forward speed being a beneficial side effect. At some point, there is no more kinetic energy to lose, and it's time to get the helicopter into the proper attitude for landing. We'll assume that the helicopter is at a very low groundspeed when the pilot decides the flare has ceased being effective and it's time to land.

The question becomes - "What pitch attitude should the helicopter have just before touchdown?".

The It-Would-Be-Nice Attitude at the end of the Flare

In a perfect world, the answer would be - the attitude that the helicopter hovers in. At the beginning of the exercise on warming up for autorotations, the student should have pointed out where the horizon crosses the windshield. This is to show the student the pitch attitude the helicopter should be in prior to landing for a zero-groundspeed touchdown. It may be an attitude that has the skids at an angle to the ground, but it is still the desired attitude. For nearly every helicopter I've flown, the hover pitch attitude is nose up, but regardless, it's the correct attitude to have at the end of the flare to get zero groundspeed.

In training, we strive to have a zero-groundspeed touchdown because we're trying to emphasize a point. But the real world is not the training world.

The Real World Correct Attitude

Let's face it, the real world is not like training. First of all, hardly any real engine failures happen over paved hard-surface runways.

Secondly, if the engine really does fail, we should be aiming for survival, not necessarily perfection.

With that in mind, the real world says that the attitude that the helicopter should be in just prior to touchdown is with the skids parallel to the ground. If you have a bit of forward speed as a result of this condition, it's not going to be too much.

If you're practicing autorotations a lot, this will also be easier on the landing gear of most helicopters - the tendency with the zero-groundspeed touchdown is: there to be enough vertical speed in the touchdown to cause long-term damage to some parts of the undercarriage. A slight forward speed (as a result of having the skids parallel to the ground) helps to judge the collective application to cushion the touchdown.

The reason for the reduced descent rate at touchdown when some forward speed is present is because as human beings we don't judge purely vertical descents well- we have eyes in the front of our head, not on the sides, and we get better depth perception with forward motion.

But there's another reason for emphasizing the "real world" skids parallel to the ground attitude. Unless judged with exquisite fineness, and absolutely zero-groundspeed is achieved on touchdown, there is likely to be some forward speed present. And the close-to-but-not absolute-zero groundspeed touchdown will almost inevitably be on the back of the skids. Couple the two things and the helicopter will rock forward while moving forward.

A very strong headwind can make things worse - the hover attitude so carefully noted earlier may have been in a different wind condition, and too slavish attention to that attitude would then have the helicopter moving rearward - something to be avoided! Putting the skids level with the ground is pretty nearly always a safe bet.

Somewhere in the Middle

The reality is that every situation is going to be different - different terrain to land on, different wind conditions, and so on. If you can get to a zero groundspeed condition, all the better. But if you can't, make sure you get to a groundspeed that is low enough to be survivable.

During the descent, fly the helicopter at a speed and rotor rpm that will enable you to make the spot. Flare as best you can and as low as possible without smacking the ground or obstructions. When the flare is going away and the landing is about to happen, use the energy in the rotor system to decrease the ground speed, if any, and level the skids if a run-on landing is expected. The attitude of the helicopter just before ground contact should be such that no backwards movement will happen, and if the surface will allow it, a smooth landing with ground speed is ideal. In a perfect world, zero ground speed is ideal and very beneficial. But it is rare than conditions are so perfect that this can be done.

Tail too Low?

There's a great deal of learning about the "personal space" the helicopter occupies during this maneuver. The location tail stinger is a great example of this. Some helicopters have a robust stinger, well placed to aid in leveling the helicopter, and are used often. Others are placed in such a manner that they are only used when really, really necessary. It's a worthwhile exercise to spend some time with a side view of the helicopter (with the appropriate skids) to figure out what is the maximum flare angle that might be used at several different heights above the ground - not that you're going to be measuring these at this critical juncture of the flight, but to give you an idea of how much space you need at different times in the flare to avoid hitting the stinger.

As an instructor, I could tell when the stinger was going to be hit.

It was often necessary to tell students that because they had the tail too low, they'd hit the stinger when they were completely unaware.

Correcting for the Inevitable Errors

Typically, the flare is the part of the maneuver that gives the most grief - either too much or too little, too soon or too late. It's not possible in a dry and dusty textbook to give a lot of advice on this -- One of the memorable sayings relevant to teaching autorotations is:.

> *"The student corrected for his/her own mistakes".*

So -- after some cyclic is used to attempt to stop the rate of descent, evaluate the results- did the control input cause the rate of descent to slow, stop or perhaps even command a climb? If the aim of the flare is kept firmly in mind, it then becomes easy to see what corrections are needed.

Too much aft cyclic stick for the airspeed results in the helicopter climbing instead of just not descending. (In other words, too much flare!).

At this point, don't continue to pull back on the cyclic stick - the helicopter will stop climbing and eventually start to descend again.

Now is the time to re-stop the rate of descent and judge what to do to salvage the situation.

Starting the flare too soon (that is, too high above the ground) will end up with the speed with respect to the ground being too low when the helicopter gets closer to the ground, and it may not be possible to stop the rate of descent, rendering the whole flare ineffective.

Too late starting the flare (too close to the ground) will require a very large nose up attitude in an attempt to stop the rate of descent, resulting in a very rushed maneuver, with a greater risk of putting the tail into the ground.

And so on.

Remember, in a real engine failure, there is no opportunity to apply power and go around! Pilots must learn to correct their mistakes and be prepared to say why they did what they did.

So What Should You Do?

Note the reaction of the helicopter in the flare: is the helicopter climbing, descending, or maintaining a constant height? If the flare is judged and carried out correctly, the helicopter will be decelerating with the rate of descent desired by the pilot (hopefully near zero).

Typically, this means the height above ground is what you want - the height depends upon the helicopter, the terrain and so on. The important part is that height is maintained as the helicopter slows (or perhaps a more exact term is the height above ground is under control). In terms of visualization - the ideal picture is for the helicopter to describe a flat or level flight path with respect to the ground.

The obvious question is how much flare is needed? The answer is just enough to stop (control) the rate of descent. Only experience will show the amount, and even then, the best that can be said is small errors will be made in every autorotation.

In the flare, the rotor RPM increases as the rate of air flowing through the disk increases - it should be permitted to increase, unless it gets too high. In a real engine failure, you might be excused if you overspeed the rotor while saving the helicopter. Being trained professionals we like to keep the rotor RPM in the green.

Flare Effectiveness

The purpose of the cyclic flare is to stop the rate of descent, but only if the helicopter has forward speed across to the ground. If the helicopter were descending at zero-airspeed on a calm day, would a cyclic flare make any difference to the rate of descent? Obviously not. If the helicopter were descending at 60 KIAS into a 60 knot wind (descending vertically with respect to the ground again) would a flare be necessary? Again, obviously not.

On a calm day, when descending at 60 KIAS, a flare is needed to stop the rate of descent, so somewhere between the extremes, the flare becomes ineffective. What to do?

First of all, make sure the ground speed and airspeed is high enough to make the flare effective. My experience has been for airspeeds less than about 40 to 45 KIAS, the cyclic flare is really ineffective in stopping the rate of descent, regardless of the helicopter type.

This is one reason for the "recommended airspeed" for autorotation found in the FM - you don't want to get too slow.

Don't get too slow prior to starting the flare.

There is a tendency to think because the ground is rushing up, we should start slowing the airspeed down. You need to overcome this.

Constant Attitude Autorotations in Chapter 17 will deal with the constant attitude autorotation.

Secondly, if you weren't able to keep the airspeed/groundspeed up, then get the skids level, aligned in the direction of flight, and use the collective lever at the opportune moment. Don't use all the collective lever at 100' AGL in a panic- the fall will hurt. Wait, with ice in your veins instead of blood, and pull when it's appropriate.

In my limited experience, judging when a flare would be ineffective has been easy, and students sort this out satisfactorily with little prompting. They still need to see the technique however.

Teaching Point

When teaching autorotations, it's necessary to debrief the student immediately after the event. With a touchdown autorotation, this means you're going to spend some time sitting on the ground with the rotor RPM at idle. With a power recovery autorotation, I'd recommend that after the helicopter is stabilized in the hover, that you land and discuss things, rather than leap off for another attempt (unless of course, you think things went very well and there is no need to debrief....

Summary of Chapter 5

After the entry, the flare is the most important part of the autorotation -- it provides the way that we stay safe, by reducing the rate of descent, and then stopping, and then landing.

Questions

1. What is the main purpose of the flare?
2. What are some beneficial side effects of the flare?
3. What can you do if the speed at the start of the flare is too low?
4. What pitch attitude should the helicopter be at the end of the flare?

Chapter 6   Energy in Autorotations

The secret to extracting the maximum flexibility from an autorotation is to understand the various energies at your disposal. Energy is the ability to do work, and the ones available in an autorotation are: potential, kinetic, and rotational. There is a subtle, but powerful interplay between these energies that we can use to our benefit -- but only if we know and understand them.

The process of getting from the time/place of the engine failure to safely on the ground can be thought of as an exercise in energy management.

That leaves us with three types of energy:

Name	Meaning	Formula	Shorthand
Potential	energy due to height above a surface	$mass \times gravity \times height$	mgh
Kinetic	energy due to motion with respect a point on the ground	$\frac{1}{2} mass \times velocity^2$	$\frac{1}{2}mv^2$
Rotational	energy of a rotating mass (the rotors)	$\frac{1}{2} Intertia\ of\ blades \times Rotor\ RPM^2$	$\frac{1}{2}I\Omega^2$

By way of further deciphering this,

I = moment of inertia (an engineering definition of the inertia of the blades) -- don't worry about it as for any one helicopter, it won't change.

(Omega) is the rotational speed of the rotor[1]

From high school physics, these energies cannot be created or destroyed, just transferred from one place to another.

Relative Sizes of the Energy

There are many ways that these energies inter-relate. Potential energy can be viewed as a source of kinetic (and rotational) energy. It's interesting to note the relative sizes of these. It's not easy to compare kinetic and potential energy, as they can be traded for one another. But the relatively small size of the rotational energy is surprising. One DVD on the subject showed that the rotational energy was a very small fraction of the combined kinetic and potential energy even at the start of a typical flare.

What makes this relative size difference important is that the rotor RPM, while the smallest energy, is far and away the most important energy -- without the rotor RPM, it is not possible to control the helicopter and all the other energies are of no use!

We've already identified three different stages to the autorotation -- the descent, the flare and landing. At each stage, energy is being converted from one type to another, until hopefully, we've wisely used all of them up. Table 6.1 describes the different sources and destinations for the energy in each phase.

---

[1] I had to put in at least one Greek symbol

	Energy Type		
Phase	Potential	Kinetic	Rotational
Steady Descent	To Kinetic and Rotational until it's nearly zero	Maintained by Potential	Maintained by Kinetic
Flare	Nearly zero, kept constant by Kinetic	To Potential and Rotational until it's nearly zero	Maintained by Kinetic
Touchdown	Zero	Zero (or very low)	Transferred to lift for cushioning touchdown

Table 1Energy Transfer in Autorotation

Since energy can be neither created nor destroyed - where did it all go? Into overcoming the drag of the rotor blades and the drag of the airframe in the descent.

The reason for discussing the transfer of energy from one type to another is that if we're going to be playing with getting more kinetic energy (energy of speed) by sacrificing height, we'd better know what the various benefits and penalties are. As you'll see, it becomes quite interesting!!

The Power of the Squared (^{2}) Term

If you remember high school mathematics, squaring a number means multiplying it by itself. Since the square term shows up in both rotational and kinetic energy terms, it's important to recognize this.

What it means is that if you have a higher speed, you have a lot more energy. Increasing the speed from 60 to 70 knots for example, means you don't have 10 more units of kinetic energy you have $(70^2)-(60^2)$ or 4900-3600 =1300 more units of energy to play with. Similarly, slowing from 60 to 50 knots means you have 3600-2500 = 1100 less units of energy to play with. The effect can be quite dramatic for a 10 knot difference in speed from what would be considered "normal".

The energy concept will be used a lot when talking about autorotation performance.

An Example of Energy

Let's take a helicopter and put it in a variety of different heights and speeds and see how much energy it has. Then let's see what can be done with that energy to arrive in some different conditions at the flare.

The equation for potential energy is PE =mass x gravity x height (shorthand- mgh).

We'll assume the mass remains constant, so we can assign it a value of 1 in our equations, and g = 32 feet per second^2.

We won't worry about the units, just the numbers.

34

Conditions at Start of Flare

A later section on performance will talk about possible trade-offs but for now, if we look at just the height above ground at the end of the descent / start of the flare using different speeds, the real picture becomes very clear.

Let's look at the difference in energy at 50' AGL (the start of the flare) with different speeds. The potential energy at 50' is 50 x 32 = 1,600 units.

What's really interesting is how much a difference in airspeed (10 knots) from 60 knots makes in the overall energy situation. At 60 knots, the kinetic energy is 3,600 units. At 70 knots, the energy is 4,900 units - 1,300 units (or 36%) greater than at 60 knots. At 50 knots it's 2,500 units - 1,100 units (or 30%) less than at 60 knots. The differences between the energy at 70 knots and at 50 knots (4,900-2,500) = 2,400 units, which is nearly 100% more than at 50 knots, or 50% less than at 70 knots. In qualitative terms, the difference is quite noticeable.

If we use 100 knots (let's say that's the maximum airspeed in autorotation) instead of 60 knots, the difference in energy at the start of the flare is (10,000-3,600) 6,400 units greater, and if we go 40 knots slower than 60, down to the ludicrously low speed of 20 knots, we have (3,600- 400) 3,200 units less energy. We can do a lot with more energy, but not a lot with very little.

Summary of Chapter 6

Understanding the types of energy that may be available to the pilot following an engine failure is essential to understanding what options you have available following the loss of the engine.

Questions

1. What are the three types of energy available after the engine fails?
2. Which type of energy is the smallest?
3. Which type of energy is the most important?
4. Does a 10 knot difference in speed (from a "normal" speed) at the start of the flare have the same effect if it's an increase or decrease in speed?

Chapter 7   Touchdown or Power Recovery?

First of all – what is a "power recovery"?

Power Recovery Defined

A power recovery autorotation terminates in a hover as opposed to landing without power. There are many reasons for ending the maneuver this way - the touchdown is the part of the maneuver with the greatest likelihood for problems; others don't want to wear out the skids; sometimes the landing area is not suitable or clearly inhospitable, and so on.

Next

The question often comes up whether it's better to carry an autorotation all the way to touchdown or do a power-recovery. The answer is an unqualified, wimpy, fence-sitting "Yes". Which doesn't answer the question at all, because the answer depends on which part of the autorotation is being taught.

Keeping the aim of the training in mind, and safety balanced with risk, the aim should be to make the training as effective as possible.

To get a really good answer to this question requires knowledge of what part of autorotations are you trying to teach, where are you trying to teach that part, and what is the student's ability to handle a lot of variables. It also requires a clear definition of what is meant by a power recovery.

It's also worth remembering that for most students, autorotations are approached with a good deal of trepidation and fear, and that there are a lot of things that happen which they may not clearly remember.

If it's not clearly remembered, then it's probably not well understood.

The previous chapter dealt with how to start off with a gentle introduction to the flare at the end of the descent. This chapter deals with the next part, the final touches.

What Part of Autorotations Are You Trying to Teach?

By this, I mean - are you trying to teach the fundamentals of airspeed control / adjustment in order to transition from the entry to autorotation up to the flare as opposed to the part from the flare to the touchdown? If the main part of the exercise is to teach only airspeed control /adjustment so as to be able to get to a particular spot on the ground, then it may be worthwhile to terminate the maneuver with a power recovery.

The student should have learned the basics of maintaining / controlling airspeed without injecting a good deal of adrenaline into the veins of both student and instructor.

Power Recovery Flare -- Now's the Time to Bring the Engine Back

At the start of the flare, apply throttle to bring the engine back into the picture of flying the helicopter. The throttle should be fully open before the collective lever is raised. The reasons for this will become clear shortly. In many helicopters, the engine coming on--line causes more handling problems for students than landings without the engine. It becomes necessary to coordinate the pedals as well - the engine introduces yaw effects absent in the real autorotation.

In the Gazelle, for example, it is necessary to use nearly full travel in the pedals (from full non-power position to full power position) at the end of a power recovery autorotation.

In many helicopters, you'll end up in a stationary hover beyond the point where you were aiming (or would have got to in a "real" engine failure). Accept this and you'll prevent overtorquing the helicopter in trying to make your intended spot. The reason is that the introduction of engine power screws up your nicely estimated control movements.

The power recovery autorotation exercises many skills, while eliminating the risks of the final part of the autorotation. If the whole exercise ends in a power--on hover instead of a no--power landing, then in "real" engine failures, the touchdown is relatively easy. Having maneuvered the helicopter to a zero groundspeed, low height condition (almost all the energy has been used) with a good amount of rotor RPM to cushion the landing, the amount of damage to the airframe and occupants should be minimal.

Power recovery autorotations to the hover have 95% of the skills necessary and less than 1% of the risk involved with actual touchdown autorotations.

Piston Engine Power Recovery

For most piston engine helicopters, changing from unpowered to powered flight is relatively simple - aside from monitoring the engine and rotor RPM to ensure that they don't go too high, it's merely a question of opening the throttle at the right time. For turbine engine helicopters, it's a slightly different story.

Turbine Engine Power Recovery

The problem for turbine engine helicopters is the governor - for those with hydro-mechanical fuel controls, the governor can only react to an error in power turbine speed from a pre-determined power turbine speed or datum RPM. For sake of argument, the datum RPM, the RPM the governor is trying to maintain is 100%. (Since the power turbine and rotor RPM are mechanically connected, we'll switch to talking about rotor RPM.). If the rotor RPM is above 100%, the fuel control will reduce the fuel in an attempt to return the rotor RPM to the correct value. The opposite is true if the rotor RPM reduces below 100% - fuel will be added. But most simple fuel controls can only measure rotor RPM, and can't anticipate power demands very well, it at all.

What's this got to do with power recovery autorotations, you might be asking? Well, it makes a large difference where the power is "recovered" in a turbine engine helicopter. If the throttle is advanced too soon, when the helicopter is in a steady descent with the rotor RPM more or less constant at the 100%, the fuel control attempts to maintain that "datum" rotor RPM. When the helicopter is flared, the rotor RPM will increase and the fuel control will reduce the fuel flow in an attempt to return the rotor RPM to the "datum" value. The compressor will be decelerating. Unfortunately, immediately after the flare, the collective will be raised to come to a hover - this puts a very large and rapid power demand on the engine, just at the time when it's decelerating. Not a happy thing! The result is an engine that is slow to respond, and a lot of dancing on the pedals and monitoring of engine gauges just at the time you should be looking outside. So when is the right time to open the throttle?

Open the Throttle at the Start of the Flare

On the other hand, if the throttle is not advanced until the start of the flare, the engine will be accelerating when the power demand is made at the end of the flare. It won't have a chance to start decelerating and the whole power recovery will be much more calm. The only issue is to remember to open the throttle!!

I would remind myself partway down by saying out loud that this was going to be a power recovery, and that the throttle would be opened at the start of the flare, terminating in the hover (or a engine off landing, as appropriate). If the student were flying, I would get a confirmation that this was understood.

Getting Back to the Hover

Once the throttle is open and the engine is on line, the power recovery autorotation is the same as the quick-stop. Taking the helicopter from the decelerating, nose up attitude to the hover is carried out in the same manner - a check on the collective lever first, followed by a small correction with the cyclic stick to get to level, some coordination with the pedals to keep straight, and the helicopter should be in a level hover.

A Bit Rushed at the End

All power recoveries are going to be a bit rushed at the end - the best comparison I can make is the fixed wing world's touch and go training. If the purpose of learning to land is to control the airplane from the beginning of the flare to completely stopped on the runway, then a lot of unlearning is done in re-applying power, re-configuring the airplane and re-accelerating to going flying again. On the other hand if the teaching point is merely the flare and touchdown, a touch-and-go might be acceptable.

The same happens in the helicopter world with a power recovery at the end of an autorotation. Learning points from earlier on in the maneuver may get lost in the rush to get the helicopter to a hover.

One way to reduce the hassle might be to have the instructor do all the power recovery portion with the student merely following through on the flight controls. Obviously this will take some prior coordination, but should have large payoff in the long run.

At the End of the First Power Recovery Descent.

The power recovery at the end of the first descent from traffic pattern height should bring you to a hover close to the desired spot, as well as show what sort of wind shears might exist between traffic pattern height and the ground. The first power-off landing should be smooth and un-rushed. If any of the following isn't what you'd expected, then perhaps some additional time spent on the basics or an appropriate exercise would be wise.

Summary of Chapter 7

A power recovery is a good way to teach nearly all of the necessary steps in an autorotation without the dangers inherent in actually touching down.

Questions
1. When should the throttle be opened in a power recovery?
2. Why should the throttle not be opened too early in the power recovery in a turbine engine helicopter?

3. Is it necessary to have the power recovery actions (throttle fully open, engine on-line) completed before starting the flare?

4. What is one way to ensure that both student and instructor remember whether the maneuver will terminate in a power recovery or an engine-off landing?

Chapter 8   Warming Up for Autorotations

Whenever I flew trips that included autorotations, I always made sure I warmed up with a specific set of exercises. Not sense messing with success.

The additional time it took to re-acquaint the various senses was always worthwhile, as it built up confidence as well as "muscle memory".

Athletes warm up for their events, and helicopter pilots should do the same. Since most helicopter pilots do not practice autorotation landings on a daily basis, it is prudent to re-align the various senses and motor reactions of the pilot. Also, since there is a great consequence for error if things aren't done correctly in training it is also prudent to re-acquaint oneself with the maneuver. I know of several autorotation training--related accidents that could be traced to a lack of warm up.

I am also aware there are those who say "real" engine failures don't give much warning, and don't permit the luxury of warming up. That is true, but since a large number of helicopter accidents happen in practicing autorotations, it stands to reason any thing that can prevent problems is worth doing. Since helicopter pilots need lots of practice in autorotations in order to develop their skills and judgment, they need to warm up.

The warm--up should be a gradual buildup, as outlined below. All of these are relatively straightforward and easy to understand, but the quickstop deserves special attention.

The sequence is the same as that used for basic instruction on autorotations, but is much more compressed. The purpose is to refresh the senses, not teach them all over again. This is worth doing on every first flight of the day, or on every student trip.

- Power On Touchdown from Hover
- Hovering Engine Failures
- Hover Taxi Engine Failures
- Quick-stop
- Steady state autorotation descent from higher-than-normal altitude
- Straight Ahead Power Recovery autorotation
- Straight Ahead Touchdown
- Make it to a (pre-nominated) Spot Straight-ahead
- Turning Autorotation
- Maximum Range Autorotation
- Minimum Range Autorotation
- Make it to a (pre-nominated) Spot - Turning.

The first power-on landing from a hover should be impeccably timed.

The "3-2-1 touchdown" should be second nature, and the skids should touch down at exactly the right second. The hovering engine failure should likewise be well timed, as should the first hover-taxi engine failure. The quick stops should be progressively more pronounced (i.e. with increasing airspeed in the "cruise" portion), and the transition back to the hover increasingly well coordinated.

Each of these items should be little more than a quick review of a previous exercise. Now we're ready to move on to putting the whole thing together.

First Warm-Up Descent

The next step in warm-up is a power recovery to the hover. It has this name as the maneuver terminates with the power on and the helicopter hovering. The main teaching points in the power recovery are the judgment in the entry point, descent and flare.

Set up the helicopter in a position where it is going to be possible to "make" the desired landing area. For the sake of argument, the flight manual says the optimum airspeed for autorotation is 60 KIAS.

At a suitable entry height above the ground and aligned with the landing area wind direction, lower the collective lever (fully down in most helicopters), put the slip ball in the middle, maintain the desired airspeed, and reduce the throttle to idle. The appropriate wording of "Practice autorotation" or "Practice engine failure" should be used.

At this point, it is worth stating that, a properly maintained (i.e. correctly rigged main rotor pitch links) helicopter will have the rotor RPM adjusted so it is within the limits set by the manufacturer.

See Rotor RPM Adjusted Properly on page 43 for more details about autorotation rotor RPM and it's adjustment.

During the descent, maintain the "normal" airspeed (the one recommended in the Flight Manual), and note how the wind affects the glide path.

Is the helicopter going to be short of the landing area or overshoot it? Make a note so the entry point can be changed for the next attempt.

At this stage of training, the teaching points are proper entry and flare techniques and the judgment for them. "Hitting the necessary spot" comes later.

One of the things to note in the descent is that at above 500' Above Ground Level (AGL), there are very few, if any, cues to the relatively rapid descent towards the ground. Secondly, the airspeed remains constant if the pitch attitude is kept the same. Note the airspeed, but don't stare at it. This checking should be a momentary glance inside to make sure the airspeed hasn't changed.

Note to instructors: one of the kindest things you can do for students in autorotations is cover up the airspeed indicator; make the student fly by attitude and looking outside.

If the rotor RPM is slightly high, then a small amount of collective lever can be applied to bring it into the proper range. A little experience will tell if you pull up too much, and there are several ways you know if the rotor RPM is too low. First, you should be able to hear it --- the sounds of the blades and transmission are different than "normal," and you should be aware of that --- hearing is an excellent sense. Get proficient at using noise to judge whether the rotor RPM is too low or too high --- then it is only necessary to glance briefly at the rotor RPM gauge --- and more time can be spent looking outside.

The other way the pilot knows if the rotor RPM is too low is the various warnings (either audio or visual). Get used to them and don't worry too much if the warning light or horn comes on.

Obviously the whole sequence need not be followed if you have an experienced student, but initially, the entire sequence should be followed until the student can show proficiency at the maneuver. And the number of autorotations to be practiced should be kept within bounds.

The student should debrief each autorotation immediately afterwards, with the instructor only correcting the student's misconceptions and / or errors.

Autorotations "En--Route"

Once the student is proficient at autorotations to the point of being able to arrive at a pre-nominated spot starting from a point within the helicopter's capabilities in the traffic pattern, then it is time to move to simulating the real world.

There are many reasons for giving student pilots practice or surprise autorotations at places and times other than the normal autorotation training areas - it teaches them to consider where they are flying (for example, flying over large areas of trees in single engine helicopters should always be avoided, unless you have at least partial knowledge of how events may transpire if the engine quits). The engine failure should be announced with "Practice Engine Failure" at the same time as rolling the throttle to idle simulates the failure - this is in order to make sure that both instructor and student know that it's not a real engine failure.

And don't be too brisk in rolling the throttle to idle.

Note to Instructors -- be cautious of giving engine failures in these conditions - despite the desire to drive home the points to your pilots. More than one practice engine failure has turned into the real thing. You don't want to have to try to extricate yourself from a self-inflicted problem!

A simulator or flight training device is an even better place for this type of training - there's very little risk of getting hurt and the instructor can inject the failure without the student being aware it's coming. See Chapter 19 for more information.

This is what the real world will be like - engine failures don't just happen when a suitable area is ideally located in front of the helicopter at the exact angle needed for a descent at the "normal" airspeed.

Summary of Chapter 8

In order to develop our skills as helicopter pilots, it's necessary to build on proper performance. Making sure that we have our senses properly tuned and developed is key to that. Doing things right every time helps to keep things that way.

Questions

1. What is the purpose of "warming up" for autorotations?
2. Should the same build-up / warm up process be used for every autorotation training sequence?
3. What is the purpose of counting down "3-2-1 touchdown"?
4. If a student is having problems with a particular segment, should the trip progress to more advanced concepts before the problem area is sorted out?
5. Should you ever make the first autorotation a power-off landing?

Chapter 9   Autorotation Performance

Before we progress to the stage of maneuvering to arrive at the desired spot on the ground, it's necessary to know enough of the theory behind what is being done with the variables of speed and rotor RPM and turning.

Rotor RPM Adjusted Properly

One of the small, but important items relating to autorotations is the proper adjustment of rotor RPM in an autorotation. Without going into lots of boring theory, it is necessary to know that the rotor RPM is correctly set by maintenance procedures. Rotor RPM in autorotation is affected mainly by weight, density altitude and blade pitch setting.

Of these, the only real variable is blade pitch setting - the pilot can adjust this using the collective, but the basic setting of blade pitch should be set properly to begin with. The proper value will normally be found in the maintenance manual and require a flight test to determine that it is correct. There will also be procedures for adjusting this basic setting in the maintenance manual - typically lengthen or shorten the pitch change rods by so much for each change in RPM desired. Typically, the rotor RPM is set for a steady descent with the collective on the bottom for a given weight and density altitude condition. See Figure 9-1.

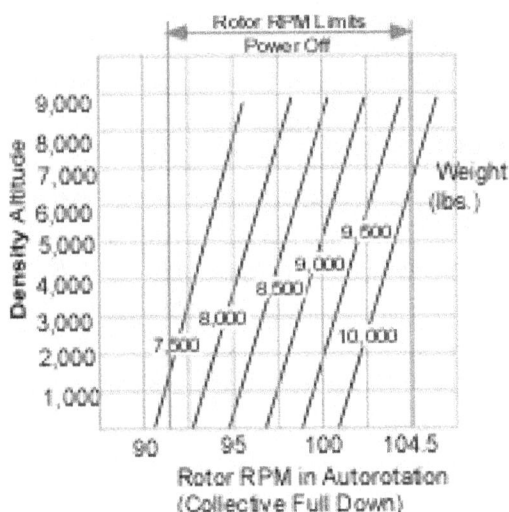

Figure 9-1 Rotor RPM Setting from Maintenance Manual

One of the points worth mentioning is that if the rotor RPM is set correctly according to the maintenance manual, but you are operating in conditions much different (in terms or weight or density altitude or air temperature) than used when setting the rotor RPM, then you will have a much different rotor RPM at flat pitch. For example, if the rotor RPM was set in a warm temperature at a medium weight, and you are now in a much colder air mass at a very light weight, the rotor RPM in a steady airspeed may be much lower than expected. With the collective on the bottom, there's not much you can do about the situation!

On the other hand, if the rotor RPM was set up on a cold day at light weight, and you are now operating at much higher density altitude at a heavier weight, the rotor RPM will be very high, and it will be necessary to use collective pitch to keep it within the approved limits.

Rate of Descent Vs. Airspeed

It is worth spending some time looking at the chart in Figure 9-2 - it contains quite a bit of information we can use.

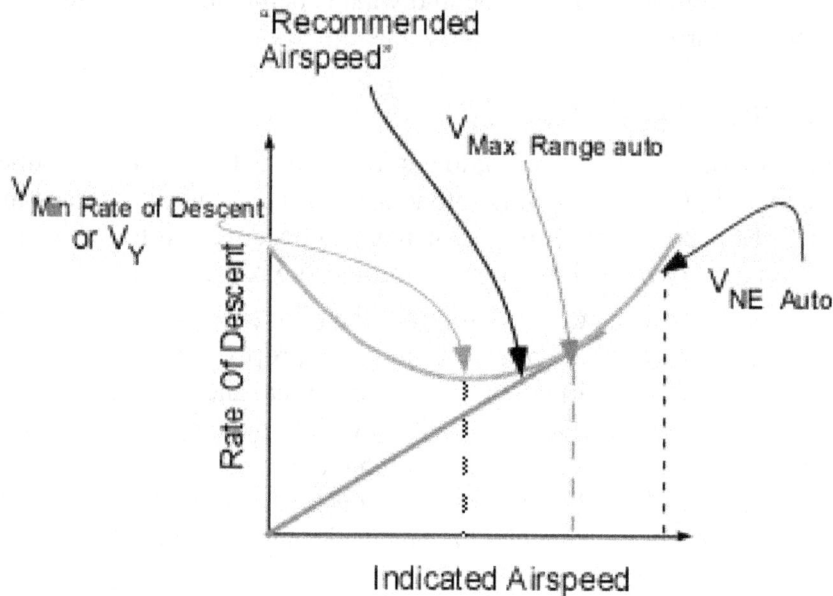

Figure 9-2 Rate of Descent Vs. Airspeed.

Where Does This ROD Chart Come From?

Flying a series of descents at different airspeeds, and measuring the rate of descent develop this chart. This provides a series of points, which are then joined to make a more general curve. The airspeed where the minimum rate of descent occurs is very close to the airspeed in powered flight where the minimum power is required to maintain level flight, and is called V_Y. We use this notation for both powered and unpowered flight, with the difference that in autorotation it provides the lowest rate of descent.

Note the "normal" airspeed for autorotations, the maximum range airspeed and the maximum permitted speed in autorotation ($V_{NE\ Auto}$).

The use of these and other airspeeds for autorotations will be covered in a moment.

The only things going for you when the engine fails are height, speed relative to the ground and rotor RPM. These need to be emphasized.

Speed relative to the ground is a combination of airspeed and wind speed. Knowing both is important. Are you upwind or downwind relative to the spot you want to land on? Since no single airspeed is going to be best in all conditions, we can ignore (within some limits of course) airspeed. In fact, one of the kindest things a helicopter flight instructor can do for a student who is learning autorotations is to cover up the airspeed indicator and make them look outside and learn to judge what's going on by using visual cues.

Some useful things can be determined from this chart. Since the vertical axis is rate of descent (height lost per unit of time), the smallest number (lowest part of the curve) will let us stay in the air the longest time. If we were only interested in time in the air, this speed might be useful, but aside from making best use of the tailwind to push us over the ground, this speed is of little use. Except that it does lead to a more useful speed - the airspeed to use to go the furtherest[1] distance over the ground (in a no-wind situation, an important point that will become clear later). It's called $V_{Max\ Range\ Auto}$, and it's derived by drawing a line from the origin (the point where the two axis are both zero) up to where it touches the curve.

For the mathematically inclined, this works because we're trying to find the fewest feet of height lost per mile traveled over the ground.

The difference between miles traveled in the air or over the ground will become important in a short while. Another way to look at this is that we want the greatest rate of change of distance with time.

If you were in descending from a great height, (for example 10,000' because you were crossing a wide stretch of water) then, maybe this new airspeed would be the one to get you to the far shore (assuming you had already passed the point where you could get back to the other shore) if there was a headwind. In our example, using maximum range speed factored for the wind gives nearly half a mile more distance from 10,000' into a 20 knot wind. On the other hand, most helicopters fly around at 1,000' AGL or less, so the difference in range possible due to wind won't be so marked. There is still one very convincing reason to use a higher than maximum range airspeed if you are uncertain if you will make the landing spot.

When the entire picture of getting to the landing spot is considered, using a higher airspeed / groundspeed has one huge and often overlooked benefit. Remember the three types of energy - potential, kinetic and rotational. At the same height above the ground, (at the end of the descent) there is very little potential energy remaining, and it is the same regardless of the airspeed used. The kinetic energy, the energy due to velocity, is definitely different between the two speeds.

Remember the squared (^{2}) term on the velocity.

Assuming nothing falls off the helicopter, the difference between the two cases just discussed is quite large. The higher speed case has the mass (a constant) multiplied by 5329 (73 x 73), while the lower speed situation has the same constant multiplied by 3969 (63 x 63). This is a difference of 34% more energy at the start of the flare! The effect is enormous. Even if the airspeed used was not the ideal one for the wind, and the glide path was marginally steeper than it should have been, the increase in energy available for flaring more than overcomes this.

The relative amounts of kinetic energy at the start of the flare, for the different airspeeds ("normal", typical maximum range and our "squeeze--everything--you--can--out--of--the--machine" autorotations) will determine the distance over the ground covered during the flare.

The numbers speak for themselves, but if you don't believe them, try the different airspeeds yourself when you are next practicing autorotations.

It is also interesting to compare the change of angle of the flight path from the descent to the flare for the two different conditions discussed above. Later, we'll compare the descent angles towards the ground using different airspeeds and rotor RPM. (If you want to look ahead, it's Figure 9-4. If the "normal" autorotation airspeed were used, the flight path in the descent would have been much steeper, and there would be less energy available to change the direction of the flight path. So two benefits come from using higher airspeeds in a wind - more energy at the beginning of the flare, and less change in the flight path angle.

---

[1] Creative spelling.

A larger change in pitch attitude results in a larger increase rotor RPM in the flare, and the higher groundspeed means we have more kinetic energy to get rid of. Assuming things were really tight, and our flight path was going to take us short of the clearing, this extra energy could be most useful.

Obviously, nothing is free - it does take some extra height to get to this higher airspeed, but the payoff is worth it.

Headwind Effect on Maximum Range in Autorotation

If we add a 20 knot headwind we get a slightly different result, seen in Figure 9-3. All the lines shift to the left slightly, but now it is possible to see the real benefit to using a higher airspeed.

Now the horizontal distance covered from 1,000' AGL at an airspeed of V_Y is only 1,800', a reduction in horizontal distance from no--wind to 20 knots of headwind is 40%, while that covered at slightly--higher--than--$V_{Max Range Auto}$ airspeed is 2,800'.

The difference between the two airspeeds (V_Y compared to the slightly--higher--than--V_Y airspeed) is close to 25%. The reason in mathematical terms for the difference between the two is that a 20 knot wind is a smaller percentage of the higher--airspeed descent (20/80) than of the slower--airspeed descent (20/60).

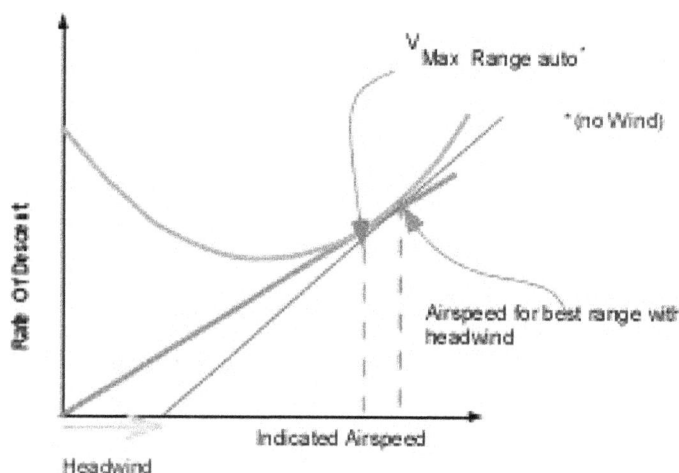

Figure 9-3 Effect of wind on Maximum Range Airspeed in Autorotation.

A Closer Look at Performance in Autorotation

It is worth spending some time looking at the Rate of Descent (ROD) vs. Airspeed chart (Figure 9-2), with the addition of a line for the maximum range airspeed in a 20 knot headwind. Note that above 60 KIAS, the slope of the curve is roughly parallel to a tangent from the origin. This means that there is not a lot of penalty in using a higher--than--maximum range airspeed.

Perhaps it is worthwhile to look at this curve in a different way - as vertical distance descended and horizontal distance traveled over the ground. Figure 9-4 has the same information as shown in Figure 9-3, but in a different form. Figure 9-4 shows the descent angles of various airspeeds, for a no-wind condition.

As can be seen, there is not a lot of difference in the glide path angle for any of the speeds. If you were only 500 feet above the ground (AGL), there might not appear to be much benefit

46

from using anything other than the normal autorotation airspeed. This ignores kinetic energy, discussed earlier. From 1,000' AGL, the distance covered using V_Y is 3,000' and that covered using $V_{max\ range\ Auto}$ is 3,600' or 600' - a 20% improvement.

And that doesn't include the added benefit of more energy in the flare.

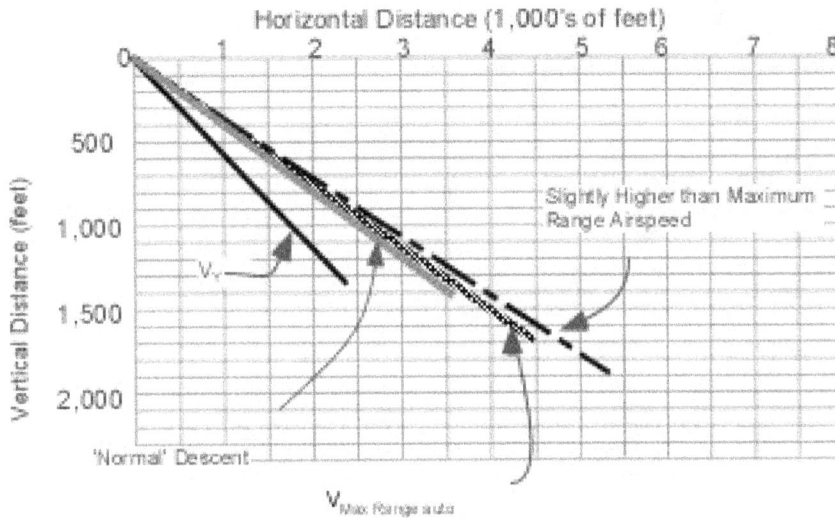

Figure 9-4 Vertical and Horizontal Distances Covered with Different Airspeeds.

Changing Rotor RPM

There is one more aspect to autorotative descent performance that needs to be addressed: the effect of reducing rotor RPM. Figure 9-5 shows how changing the rotor RPM at the same airspeed can change the rate of descent quite markedly. If we consider the "normal" rotor RPM for this helicopter to be 480 RPM, it can be seen that reducing the rotor RPM to the minimum power-off RPM reduces the rate of descent by about 200 feet per minute. Conversely, allowing the rotor RPM to increase to the maximum power-off level, increases the rate of descent by about 200 feet per minute. Rotor efficiency has a lot to do with this, which is a subject best left to more serious books on helicopter aerodynamics by experts like Ray Prouty, so we'll just accept that this is the way things are for now.

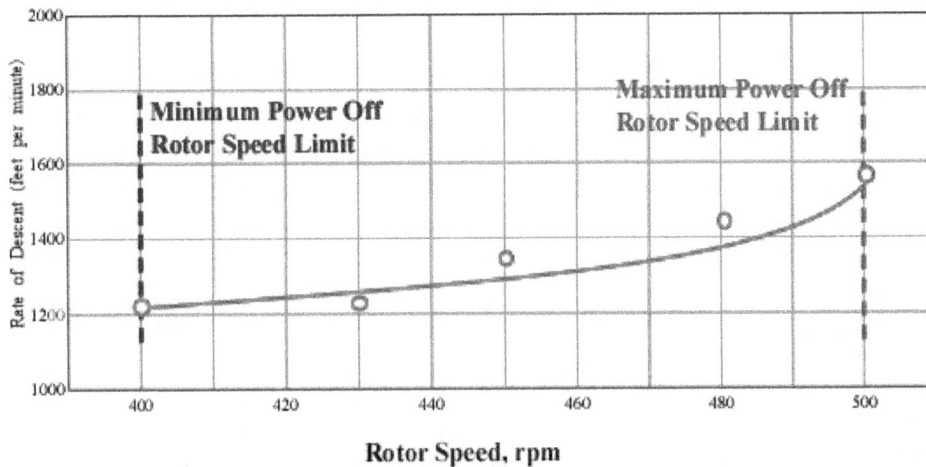

Figure 9-5 Rotor RPM And Rate of Descent.

How does all this affect us? Figure 9-6 shows the change in glide path angle resulting from using minimum power off-rotor RPM during the descent compared to the "normal" RPM. If the autorotation is started at 1,000' above ground, the helicopter will travel a horizontal distance covered using "normal" rotor RPM (line A) is 2,800'. Using minimum power--off rotor RPM (line B), the horizontal distance covered is 3,800', a difference of nearly a quarter mile, or 35% farther!! 1,000' is a large distance and will significantly increase the possible landing areas available.

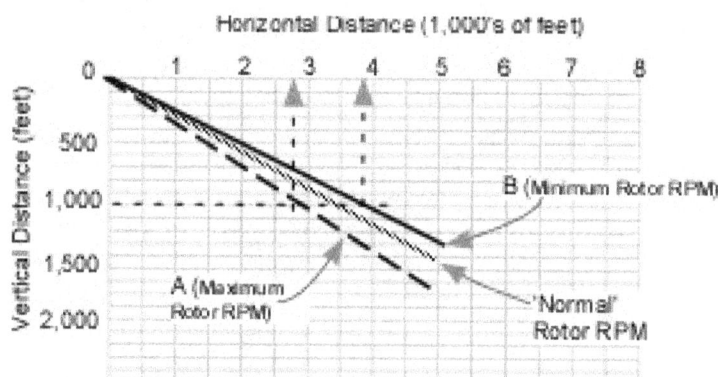

Figure 9-6 Effect of rotor RPM on Glide Path Angle (and distance traveled across the ground)

To demonstrate the cumulative effect these variations can have, let's assume we start our engine failure from 1,050 feet above the ground, with a 20 knot headwind. We'll start our flare at 50' above the ground, so we have 1,000' to descend.

Let's say that we're in autorotation at our "recommended airspeed" of 60 knots, which would normally have a rate of descent of 1400 feet per minute, but we allow the rotor RPM to increase to it's maximum power off value (we just lower the collective and don't pay too close attention) - the effect is an increase in rate of descent of 200' per minute, so our rate of descent is now 1,600 feet per minute. We have (1,000/1,600) or 37 seconds in the descent. The distance across the ground we'd cover would be (40 knots time 6,080 feet per nautical mile times 37 seconds divided by 3,600 seconds per hour) = 2,500 feet over the ground.

Now, let's consider that we use 80 knots airspeed, which would give us a rate of descent of 2,000 feet per minute at "normal" rotor RPM, but we'll reduce the rotor RPM to the minimum power off rotor RPM.

Note that we can't use the rate of descent values in Figure 9-5, because we're using an airspeed that is above the normal $V_{Max Range Auto}$ airspeed. We'll assume that the effect of reducing the rotor RPM to minimum is to reduce the rate of descent by 200 feet per minutes as shown in Figure 9-5. Now we have a rate of descent of 1,800 feet per minute, which means we'll be in the descent for a mere 33 seconds (5 seconds shorter than if we use the "recommended" airspeed for autorotation). However, this time, we cover (60 knots groundspeed times 6,080 feet per nautical mile times 33 seconds divided by 3,600 seconds per hour) = 3,344 feet over the ground.

That's a difference of 800 feet or nearly 30% increase in distance over the "recommended" airspeed. This is shown in Figure 9-6.

If you had to cover at least 3,000 feet to get to a suitable landing spot, would you know how to do it?

Both of these previous examples have been taken in isolation. In a real autorotation, the ride doesn't end with a constant airspeed landing - all that airspeed has to be converted into another type of energy, and this is where the real secrets of getting the most from autorotations lies.

Kinetic Energy in the Flare

To keep things simple, in the beginning of this discussion, we will deal only with the no wind situation, so it should be normal to consider using only the advertised "maximum range" airspeed in autorotation.

It should be normal, but this is not a normal book and we are interested in the whole picture. The published V_Y is the best airspeed for maximum range only if we were to ignore the energy available at the start of the flare. For this demonstration we want to use the maximum speed possible in autorotation, just as we used the minimum speed possible in demonstrating how to "get rid of" range.

For sake of argument, let's say we have already decided to use the highest speed possible, with the rotor RPM at the low end of the permitted power--off range.

If we continue with the comparison that was developed above, if we use the "recommended" airspeed, when we arrive at the flare, our groundspeed is only 40 knots, whereas if we use the higher than $V_{Max Range Auto}$ airspeed, we have 60 knots of groundspeed when we start the flare.

The difference in kinetic energy is huge (relatively) - 1600 "units" for the 40 knot case compared to 3,600 units for the higher speed.

That's a difference of 125%.

Why all this emphasis on the higher speed? In summary -

Published $V_{max range Auto}$ is the best range airspeed **only for a no wind situation**.

Tailwinds

The main discussion in this chapter has been about how to get the most range when autorotating in a headwind. Obviously, life can sometime give us a tailwind. Just as the over-riding reason for increasing airspeed into a headwind is to reduce the amount of time the headwind affects us, for a tailwind, we want to maximize the time it affects us. So, in the case of a tailwind, reduce the airspeed to no slower than V_Y and reduce the rotor RPM to the minimum power-off rotor RPM.

Something for Nothing

It appears to be counter-intuitive that starting from the same height by reducing rotor RPM to the minimum power off RPM, and using a higher than normal airspeed you can go farther, and end up with more energy at the start of the flare. Hopefully the theory is clear. Here's how to prove it to yourself.

Proving the Theory for Yourself

On a day with a moderate, but steady wind (10-15mph), set up for an autorotation from the cruise. Note the entry location. After simulating the engine failure, and getting the cyclic back and the collective down, set up the airspeed for a "normal" autorotation and using "normal" rotor RPM. Maintain this till the "normal" flare height, and then flare. Note where you end up at the end of the flare (whether in a power recovery or touchdown).

Next, go back to the same entry conditions of cruise airspeed and height above the ground and location over the ground and repeat, this time using $V_{\text{Max Range Auto}}$ and then reduce the rotor RPM to the minimum power-off rotor RPM. Maintain this till the "normal" flare height, and then flare. Note where you end up at the end of the flare (whether in a power recovery or touchdown).

Convinced?

Summary of Chapter 9

The purpose of this chapter is to show how to get the maximum possible distance from the kinetic energy given by height. The effect of headwinds and tailwinds is clearly demonstrated.

Questions

1. What are the advantages of using a higher than normal airspeed in an autorotational descent?
2. What wind conditions is the $V_{\text{Max Range Auto}}$ autorotation airspeed shown in the flight manual good for?
3. What airspeed should you use if you have to autorotate into a headwind and you need to cover a lot of ground?
4. What is the slowest airspeed should you use if you have to autorotate with a tailwind?

Chapter 10   Putting it all Together

In this chapter, you will learn how to use all these techniques to vary your range in autorotations so you can arrive at a spot of your choice. (Within reason and the physical ability to get there, of course.).

During the descent you must check the temperatures and pressures are in the normal operating range and that the engine is still operating and will be available for you to commence your power recovery.

Judging Wind Direction from Naturally Occurring Cues

It's assumed that you know how to judge the wind direction from cues such as wave patterns on bodies of water, flags, smoke / steam plumes, and the like. If not, see "Cyclic and Collective" for more information.

Cone of Possible Areas

The potential and kinetic energies available can be used in many ways (some positive, some not so good)- the point is that they can be used get the helicopter to a wide variety of places following an engine failure. The problem is how to describe this area. The best description that comes to mind is a cone extending down and around from the helicopter, like the beam of a landing light. Within reason, it should be possible to get to any suitable spot within this area, including an area behind you. This cone is not valid for areas close to the ground, especially near the height velocity curve, but for a helicopter in a medium--altitude cruise, it is a useful concept. An example of such a cone is shown in Figure 10-1.

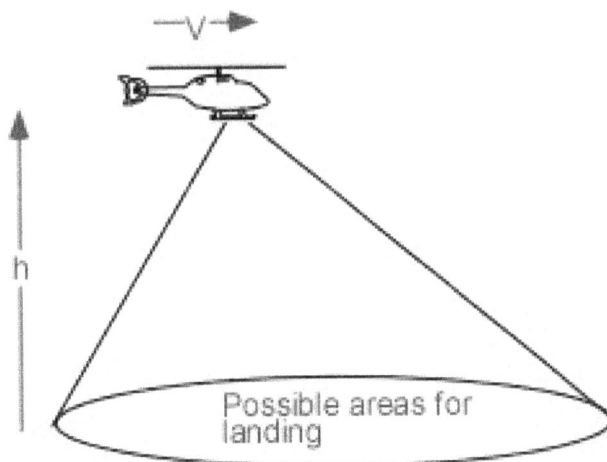

Figure 10-1 Cone of Possible Landing Areas

If a wind is added, it only shifts the shape of the cone slightly, as shown in Figure 10-2.

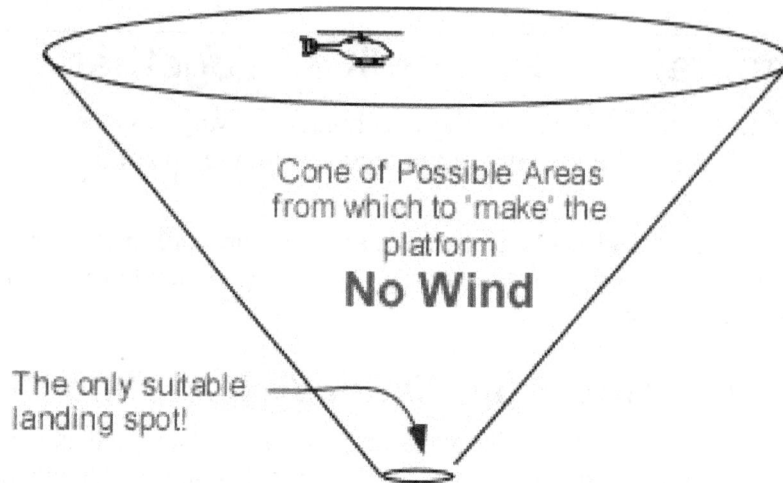

Figure 10-2 Cone of Possible areas No Wind

Obviously, we can never predict when the engine will fail, and so we can't always be in the ideal position (as in training). The next section discusses the variations the successfully-- trained pilot can handle to get to the necessary spot.

Reverse Cone of Energy

Real life hasn't yet paved over, let alone smoothed out the whole world so we have to be able to find somewhere suitable to land,

Suppose you are flying over a heavily--forested area, and there is only one suitable clearing you can land in. Wisely, you have placed yourself in a position where you think you can make it to the clearing.

The concept of the "cone" can still be applied, but in reverse. The pilot must be able to weigh the options to arrive at the landing site.

Figure 10-2 shows this reversed or ground-based cone (i.e. the possible places you could be to make this spot) for a no-wind situation.

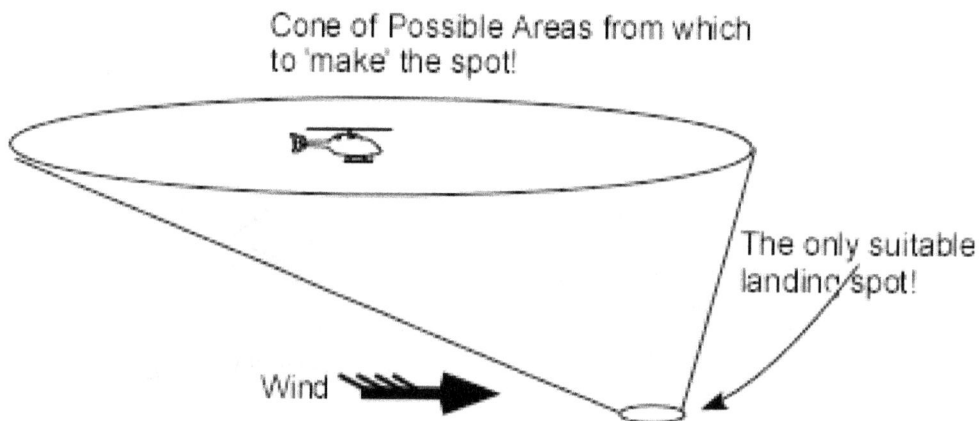

Figure 10-3 Reversed Cone (with wind) - Possible Volume.

On a no-wind day, a cone centered on the landing site defines the volume of the air within which it is possible to be, and make a safe landing in the clearing. (The cone is much larger than you might expect!)

Figure 10-3 shows it with wind effects considered. Think of this as a giant funnel leading to the desired landing spot.

Landing Site is Straight Ahead and an Ideal Distance Away

Let's start with the clearing directly in front. You must coordinate the controls in a timely fashion to arrive in the clearing. How do you go about it?

Energy management starts to come into play now - you have the cruise airspeed (kinetic energy), the rotor RPM (rotational energy) and the height above the clearing (potential energy) to play with. How should we use them?

At this stage, you are using up potential energy, and it can go into only two other forms of energy; rotor RPM (rotational energy) or airspeed (kinetic energy).

Be Into the Wind for Zero (or Low) Groundspeed

Obviously you wouldn't want to get the air coming at you from behind, so a good piece of advice is to make sure you know the approximate direction of the wind at altitude before you try this.

The first time I ever saw this demonstrated, the instructor had placed me very high and close to the nominated field, and I was clearly not going to make a good show of things. He suggested that I turn into the wind, (which put the landing area out the right window) and slow to zero (or low) groundspeed. When things looked about "right", a combined turn and dive towards the area sorted things out beautifully.

I can still clearly remember the "mind expansion" as that unfolded.

Why not do Turns instead ?

There is a school of thought that might advocate using turns to get rid of too much height. I don't recommend it because of the rapidly changing sight picture - the desired landing spot is moving all over the windshield, winds that are not exactly "right" can change things significantly, and there is the need to raise and lower the collective to keep the rotor RPM in its proper place. So, to simplify things, I don't recommend turning unless you are out of options.

Why Try the Variations?

The two main variations in the airspeed, namely low speed and higher-than-maximum-range airspeed, will show the two extremes of how to get to the clearing.

It should start to develop your ability to judge the techniques to get you safely to the place you want to land.

All of these variations have been merely to get us to the end of the descent, and the next stage in the autorotation, namely the flare and landing are no different from what has been covered earlier.

Prove It!

In order to prove this to doubting students (and sometimes experienced pilots), it was necessary to enter autorotation from the same start conditions and place over the ground. Initially, the "normal" method of slowing to the desired airspeed and maintaining rotor RPM at the "normal" RPM was used, and a note made of how far over the ground the helicopter traveled. The next entry, from the same height and speed and over the same spot on the ground was to keep the airspeed quite high (even above $V_{Max\ Range\ Auto}$ and to reduce the rotor RPM to the minimum power-off RPM. The distance traveled was so significantly greater that no-one ever doubted the benefits.

With not much work, you can do this as well.

Applying the Energy Analogy to Slower Airspeeds

It is worth noting that higher speed will provide more energy in the start of the flare, the same logic can also be applied to slower speeds - energy available for the flare drops off rapidly as speed decreases.

It takes energy to change the flight path, and when there isn't enough of it, the rate of descent can't be stopped by using kinetic energy (that is, by cyclic flaring). Taken to a ridiculous extreme, if you were descending in a low--airspeed autorotation, flaring with cyclic wouldn't change the flight path at all. Surprisingly, at speeds less than about 40 KIAS, flaring won't change the flight path very much either- there isn't enough kinetic energy. But we digress - back to the variations...

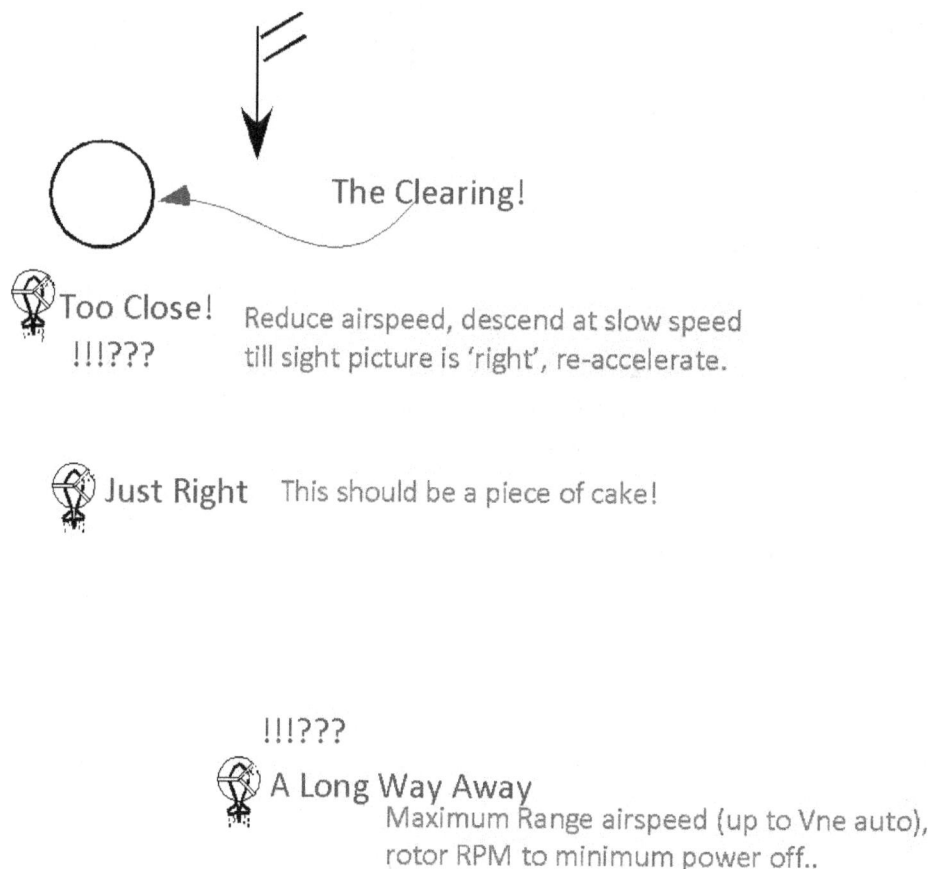

The Clearing!

Too Close!
!!!???   Reduce airspeed, descend at slow speed
                  till sight picture is 'right', re-accelerate.

Just Right   This should be a piece of cake!

!!!???
A Long Way Away
                  Maximum Range airspeed (up to Vne auto),
                  rotor RPM to minimum power off..

Figure 10-4 Possible Locations with Respect to Landing Site - Into wind.

Situations with Respect to the Landing Spot – Into Wind

We'll state with the landing area is directly in front of us while we're going into wind.

There are three basic locations you could be in respect to the clearing.

Ideally situated for the autorotations seen in training, too close to it, or too far away. Figure 10-4 shows these three locations. The ideal situation would require nothing more than lowering the collective and setting the airspeed to the one your instructor drilled into you and waiting for the exact moment to flare. Seldom are we so lucky.

Too Close to the Landing Site!

An embarrassment! How could you possibly say to anyone, (let alone another helicopter pilot) "I couldn't make it into the clearing, it was too close"?

In this case we must get rid of some of the energy, but in a way that will allow the maximum number of options - no sense throwing anything away. A 360° turn (or perhaps even less) could be carried out here. I don't like turns in autorotation as it becomes difficult to judge closure rates (especially if there is a wind), the rotor RPM needs constant attention, and so on. Also, the wind changes with altitude, and makes life more difficult to judge things. At some point in the turn you have both turned your back on the landing site and are going away from it- not good things. Personally, I wouldn't turn.

My preferred option is to change airspeed - decelerate the helicopter to a slow speed while keeping the clearing in front, or at least off to the pilot's side. A zero speed descent with the clearing out the same side as the pilot is sitting means the sight picture doesn't change, and only a small, short--radius turn is needed to line up again. When the time is correct (and only judgment and experience show when this is), the nose is smartly lowered and the helicopter accelerates into a normal autorotation profile. This whole maneuver is just a variation on a low speed autorotation.

Once in the flare, the rest of the maneuver is standard.

Landing Site Far Away

A more difficult predicament is the "going for range" situation. The $V_{NE\ Auto}$ was developed for a good reason – bad things happen if go faster - in one light helicopter, the rotor RPM decays rapidly. Between the normal airspeed for autorotation and $V_{NE\ Auto}$ there is quite a range of speeds to choose from.

At the moment of truth how should you judge which airspeed to use? Since you have no way of knowing the wind between you and the landing spot, and time won't permit you to get out the graph of Rate of Descent vs. Airspeed anyway, and you can't measure either the height above ground or the distance to the landing spot, you are left with the Mk1 eyeballs you were born with and trained judgment. The way to get the judgment is to see the tradeoffs possible in real life (or a very good simulator, which is rare!).

There is no always--correct answer - but a good rule of thumb is

Undershooting? Go faster.

Overshooting? Go slower.

This may seem incorrect, but look at the situation this way. Going faster temporarily increases your rate of descent and your descent angle becomes steeper, but that's not the whole story. For one thing, you'll reduce the time you're exposed to a headwind. For another, the proof lies at the start of the flare. You will have a much higher airspeed/groundspeed at the start of the flare, with the attendant increase in energy.

Nearly On Top of the Landing Site and Too High

If you're nearly on top of the landing site, and too high, the issue is how to get rid of the height. As you decelerate and maintain altitude, turn to put the wind on the side you're looking out of. Slow the forward speed to something you're comfortable with (you can go to zero airspeed, but you won't be able to measure it, so just go to a low airspeed).

When the sight picture looks "normal", lower the nose positively and re-acquire the normal autorotation profile. The rest is "standard".

Other Situations with Respect to the Landing Site

What if the landing site is just behind, or off to one side, or downwind? The only change this makes is that now we must turn to get into a good landing position - all other things remain the same. In all of the situations in Figure 10-5, the helicopter is considered to be heading downwind.

Landing Site a Long Way Ahead, but We're Going Downwind

Obviously, the first thing to do is to get close to the landing site - maximum range airspeed (or higher, up to $V_{NE\ Auto}$) and minimum power-off rotor RPM till you know you're going to make the spot. Then, when you get close and have to turn into wind, make the turn quite aggressively, and you can flare while turning. Nothing says the flare has to be done in wings level flight.

Clearing is directly Alongside, but We're Still Going Downwind

If the clearing is directly alongside, then it is a matter of establishing the helicopter in autorotation and simultaneously rolling in to a turn, and judging the rate of turn to roll out nicely lined up and into wind. Two important things to consider here are firstly, the rotor RPM increases during the turn so it may be necessary to hold a small amount of up collective. The second point is to not lose sight of the landing area - keep looking at it, keep judging the closure rate, turn rate and so on. Nothing says the flare can't be made in the turn. Since the turn took you into wind, things worked out really well. We should all be so lucky!

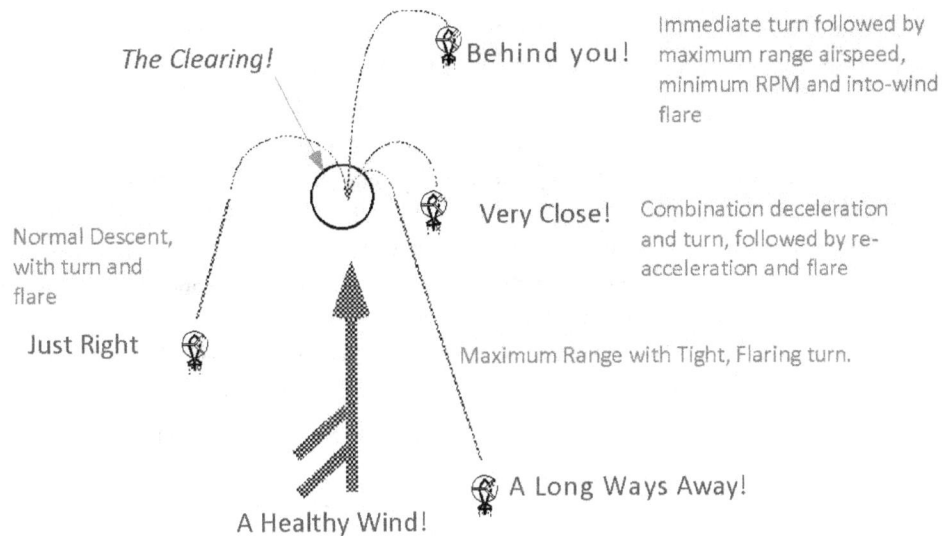

The Clearing!

Behind you!

Immediate turn followed by maximum range airspeed, minimum RPM and into-wind flare

Normal Descent, with turn and flare

Very Close!

Combination deceleration and turn, followed by re-acceleration and flare

Just Right

Maximum Range with Tight, Flaring turn.

A Healthy Wind!

A Long Ways Away!

Figure 10-5 Other Possible positions with Respect to the Landing Site.

Landing Site is a Long Way Behind, Helicopter Headed Downwind

First the good news - you don't need a long memory. From 500' AGL, anything passed behind more than 5 - 7 seconds ago is too far away.

If you're within this short time frame, the first thing to do is turn - and turn hard. Roll on the bank angle, and pull the nose around while raising the collective slightly. Don't worry about raising the collective too much, the low rotor horn tells you very soon, and just lower the collective slightly to make the horrible noise go away.

The greater the amount of G pulled in the turn, the more the rotor RPM tends to increase, so judge collective application accordingly.

Keep the airspeed up, and as you roll out of the turn, lower the collective and try for maximum airspeed and minimum rotor RPM, for all the reasons discussed before.

It sounds very easy to write this, and hopefully to read it in the comfort and safety of a chair. If you are reading this while trying your first autorotation - put the book down now and look outside!

Of course, things won't be this cut and dried when you do it for real, but understanding the why and wherefore will help you concentrate on the flying outside.

Combinations!

Nearly any other situation with regard to a clearing is going to be a variation of one or more of the scenarios given above. If you can handle these variations on the theme about autorotations, you should be able to handle anything when the engine fails.

Summary of Chapter 10

This chapter has covered a lot of material and brought a lot of theory into sharp focus. There are only so many different places you can be with respect to a suitable landing site, and if you know the variations in technique, there is more likelihood of reaching the necessary place!

It's interesting to note that some countries have specific exercises on these variations as part of the training syllabus for a helicopter rating. Wish they all did.

Questions

1. If the only available landing site is a long way ahead of you, and you're downwind, and the landing site is small and surrounded by tall trees, describe what you'd do and where you'd start to turn.

2. What is the more important to maintain - rotor RPM or airspeed?

3. You've just passed an ideal landing spot when the engine fails. There doesn't appear to be any others in range. You're into wind at the moment. What would you do?

4. Is it necessary to be lined up and wings (rotor disk / fuselage) level by 100' AGL?

5. Describe how a headwind will affect the area that can be reached in autorotation.

6. Describe how a tailwind might affect your choices of possible areas if you have an engine failure in the cruise from 1,000 AGL.

Chapter 11   Pass on the Grass

Touring a helicopter training center (no names!) I saw a number of beat up airframes sitting under tarps. Asking about these, I was informed they were all due to autorotation accidents. Further questions led me to the fact that they were all as the result of autorotations done to the grass... A sign of slight exasperation must have escaped my lips - even in my limited sphere of knowledge I knew that these could have been prevented if the landing surface had been a paved or concrete surface.

Autorotations are tricky enough maneuvers, with much more than just rotor inertia playing a part in making them easy to complete for the student, and keeping the instructor's reputation and stress level safe.

The reason for teaching autorotations to students is primarily to show them how to survive an engine failure in a single engine helicopter.

There are other reasons, but this is the main one. The concept that should be stressed for the bottom of the autorotation is how to get the machine's vertical and forward speed down to something survivable.

Beyond that, we're building student confidence and judgment.

Whenever we do training, we need to make sure we stack the odds in our favor. For most of the autorotation this is normal - we start from a place we know we can make a safe landing (very few will simulate engine failures over the middle of a forest, for example). We build up to the event with power recoveries and other maneuvers. We don't really fail the engine. And so on.

First of all, if you can guarantee that all your touchdowns will be zero speed, you can safely skip this article. On the other hand, if groundspeed may happen, read on!

Let's look at the reasons why the touchdowns shouldn't be done to grass. There are three reasons that I know of.

The Three Reasons

Landing Gear Design

The first reason is the landing gear on skid-equipped helicopters is designed to spread laterally under load. This is how the gear was designed and when drop tested, it is allowed to spread. If the landing is on a firm, (relatively) hard surface, the gear can and will spread.

If landing on a soft surface, the skid tubes dig a small depression in the surface and are prevented from moving sideways. The stresses get transferred to other parts of the landing gear - parts other than the original design considered. For one helicopter, it's easy to tell if it's done a lot of autorotations to the grass, as the tail stinger sits lower than normal due to the cross tube being bent. (it can eventually fail because of this unusual loading).

Hard Surface   Soft Ground

Cross Tube can flex   Cross tube can only
up / down and   flex up and down
sideways

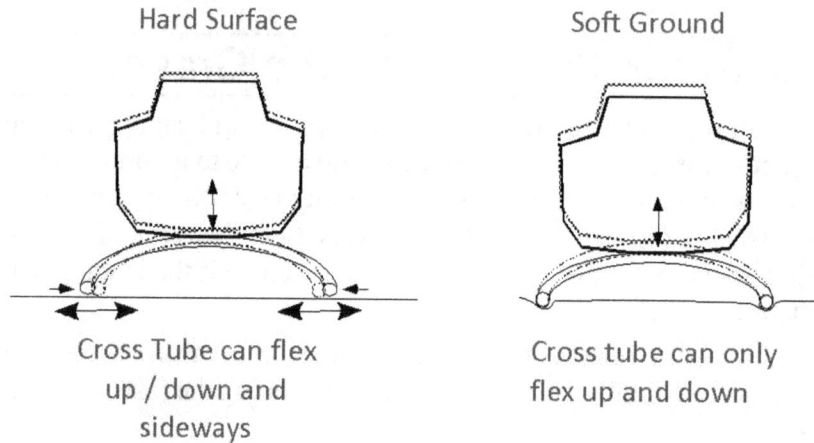

Figure 11-1 Landing Gear Movement during Heavy landing.

Ground Friction

The second reason is that the friction of landing on a soft surface is greater than a paved hard surface. Seems strange until you consider the amount of skid tube area that is in contact with the ground. On a hard surface, it's only the bottom of the skid (and if you have the heavy-duty skid shoes, even less than that) touching the ground.

In soft ground, the sides and the bottom of the skid tubes are in contact with something trying to slow you down. This might not be too bad, except that most helicopters have a high center of gravity.

The combination of the high center of gravity and the greater friction tend to want to slow the helicopter down quite dramatically. Too dramatically, and the helicopter may nose over and then fall back. The nosing over, combined with low rotor RPM can result in blades attempting to shorten the tail boom since the natural reaction of most pilots is to pull the cyclic back when the nose goes down without asking. (Narrowly avoided that trap once…).

Related to the surface area in contact with the skids is the often hidden inconsistent nature of nearly any soft surface. Drive across it slowly in a car or bicycle and feel the bumps. If that's not bad enough, there will be different degrees of hardness - and an unexpectedly soft bit may cause the skid to dig in and slow abruptly. Maybe stop more abruptly on just one side. Paved or concrete surfaces that are in good condition have a remarkable consistency and smoothness in comparison to nearly any soft surface. And if they're not in good condition, it's pretty easy to see it's not smooth whereas grass can often hide uneven ground.

The reason this uneven surface is an issue is that the vertical position of the CG in most helicopters is very high, and a strong friction force on the ground working on a high CG means that not only will the deceleration be strong, but that it will try to tip the helicopter nose forward. Since you're typically at a low rotor RPM at this stage of the autorotation, there isn't much restoring nose-up force available from the main rotor even if you did move the cyclic aft. Aft cyclic will only make you feel good (briefly) but not stop the nose down motion. It will also probably result in shortening the tail boom as it is rising up while the main rotor is flexing down.

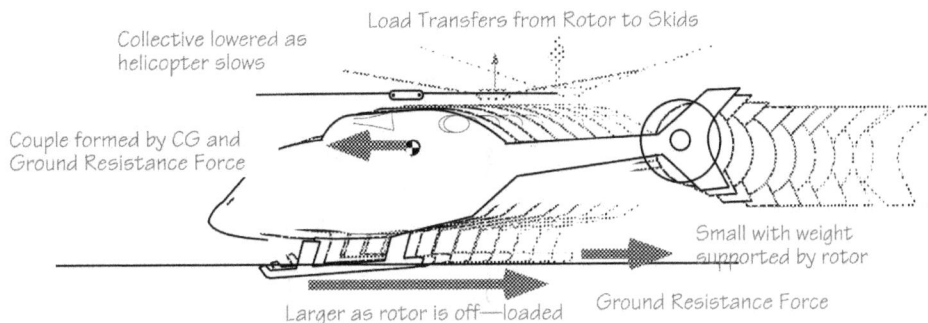

Collective lowered as helicopter slows

Load Transfers from Rotor to Skids

Couple formed by CG and Ground Resistance Force

Small with weight supported by rotor

Larger as rotor is off—loaded

Ground Resistance Force

Figure 11-2 Sudden Deceleration.

Pointing vs. Traveling

The third main reason is related to the difference between where the airframe is pointing and the direction it's traveling. When you touch down with forward speed on a soft surface a complex set of physical laws makes sure that the direction of travel and pointing are the same. For example if you're pointing left of the direction of travel and the right rear skid touches down first, on soft ground, a very strong couple will result from the friction force behind and to the right the CG. This will try to twist the airframe so these two forces line up. If the left skid touches shortly after, this will add a friction force on the other side and the pair of friction forces will try to make sure the CG lies between them. Looking at this from above is scary enough, but remember that the CG of the helicopter is also high.

Enough of a difference between pointing and travel could roll the helicopter over.

On a hard surface, the friction is much less and the tendency to want to line everything up is much less. Many times I've slid along the runway pointing in a different direction than the way the helicopter was heading.

I don't want to pick on the venerable Bell 206 series, as it's one of my favorite helicopters for autorotations, however the transmission on these machines has a lot of lateral movement, and when the direction of travel and direction of pointing don't agree when landing on a soft surface, the effect on these machines is a phenomena called pylon whirl. The pylon holding the transmission is free to move laterally, and normally this is not a problem as the rotor has a lot of inertia that keeps things lined up. However if the fuselage is pointing in one direction and the helicopter is actually moving in another direction on landing with low rotor RPM, the jolt can displace the transmission / pylon assembly to the point where it starts to move side to side quite violently.

The pylon is rocked pretty violently sideways with respect to the fuselage and can actually hit the cowlings. Never heard of it happening in an autorotation to a hard surface.

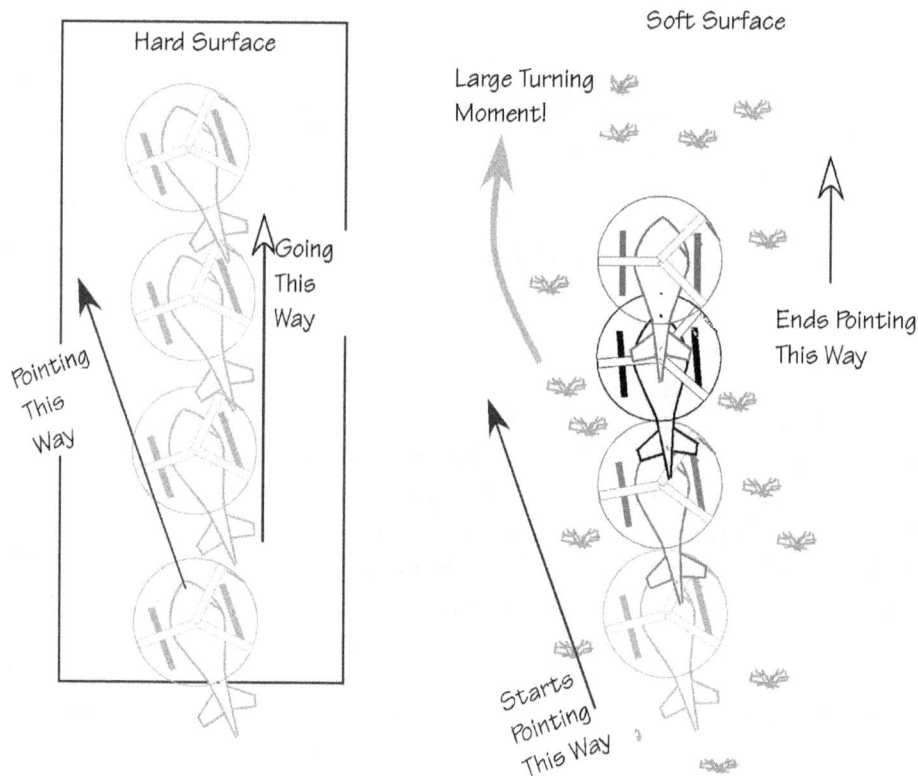

Figure 11-3 Pointing vs. Traveling.

"Problems" with Autorotations to Concrete

Most of the objections that people raise about doing autorotations to concrete are only perceptions. But here they are anyway.

"It feels rougher in the seat of the pants." The airframe is designed for something other than how things feel in the seat of your pants, and the decelerations and lack of twist on the airframe is in fact easier on the airframe.

"There is a lot more noise than doing it to grass." Get used to it.

"It wears out skids more quickly." Most skids are fitted with replaceable shoes. The solution is to use heavy -duty skid shoes made from tungsten carbide or some other very hard substance. They weigh a few pounds more than normal skid shoes, but last for a very long time. (and the extra weight will move the vertical position of the CG down a fraction of an inch or so).

By the way, if you think that normal skid shoes last a long time when landing on grass, you might think again. More than one person has been surprised when a skid has broken and found to have been slowly wearing away due to autorotations to grass surfaces.

"Skid shoes tear up the runway." This myth might be an issue (and even that isn't proven) on very soft asphalt in warm weather, when normal skids would also dig in. Look for a concrete or hard pavement surface. If you go and run your hands over the scratches made on these surfaces by heavy duty skid shoes, you can't feel the difference - what's left on the surface is actually the skid shoe material, and there is no gouging of the concrete.

A word of caution about paved surfaces - try to keep the skids off of any painted runway markings - not because you'll wear them off, but because they have a different sliptivity[1] coefficient and that may cause minor problems with maintaining alignment.

For concrete surfaces with tar joints, make sure you don't get one skid stuck in the tar joint - it's pretty sticky stuff in warm weather.

One minor point that also needs clarification - it's OK to do hovering engine failures to the grass - there is almost no motion or twisting, and the vertical speeds are normally quite low.

More autorotation accidents happen on grass than on paved runways.

Bell Helicopter and Robinson Helicopters do autorotations exclusively to paved hard surfaces. So does the US Army. Why doesn't everybody?

A Novel Way to Solve the Problem

One training organization I know of that had no option but to carry out autorotations to the grass solved their problems by ensuring the grass area was always well drained, the grass was kept short (to show up any uneven spots), the area was rolled to keep it flat, and to top it all off, always kept full-length skis on the helicopters that were doing autorotations - even in the summertime.

Summary of Chapter 11

This chapter presents the reasons why touchdown autorotations should be done to hardened, prepared surfaces, and not to grass or other soft surfaces.

Questions

1. What are three reasons for not doing autorotative landings on grass surfaces?
2. If you have no option but to carry out an engine --off landing on a grass surface, what can you do to minimize the risks?
3. Do special skid shoes damage asphalt or concrete surfaces?

---

[1] I like to make up my own words, but this one I inherited from my father.

Chapter 12   Practice vs. Real Autorotations in Turbine Helicopters

Fortunately, not many of us have real engine failures. And we should be practiced enough to know how to recognize and deal with one if it happens. So, it should be no big surprise when the engine fails - instinctive reactions take over, and we do what should come naturally.

Except, in some turbine engine helicopters, what we've practiced and learned to judge things by, is not always what happens.

How is this so?

How Turbine Engines Really Fail

One unrealistic difference between training and reality is how a turbine engine failure is simulated compared to how it really fails. For nearly every turbine helicopter I've flown, retarding the throttle to idle is the accepted way to simulate the failure. But engines don't quietly run down like this all the time. For one popular engine, with 6 axial stages in the compressor, the deceleration (regardless of how fast you reduce the throttle) is decidedly gentle - the fuel control can't reduce the fuel too quickly or the engine will flame out (and we wouldn't want that now, would we?). Even if this engine suddenly lost fuel, the deceleration of the compressor would be slow because of the mass of the parts. While the compressor is still compressing, air under pressure is getting to the power turbine, making the deceleration of the rotor slower than it would be if, say, the power turbine blew up. A different model of this engine has a single centrifugal compressor, and the rate of reduction in power when the throttle is retarded in that engine is quite dramatically different.

Other things can cause the engine to stop producing useful power - the drive shaft to the transmission can break or the over-running clutch can fail. Rare, but these things have happened. So, another thing to make the simulation of an engine failure unrealistic is the rate that the engine power goes away compared to the real thing. For instance, a break of the system between the engine and the main-rotor transmission will cause the power to go from normal power to zero in an instant.

Only in a simulator can some of these items be safely shown, and in a real helicopter, the price we pay to simulate things safely is a bit of unrealism.

But the next difference is one that has caught a lot of people by surprise - and it's a nasty surprise.

I must stress that this phenomena does not happen on all turbine engine helicopters.

Even with the engine at idle, the power turbine is still producing power. If you don't believe that, then how does the rotor stays turning with the engine at idle on the ground? (Fixed shaft engines as fitted to the Gazelle and the Alouette / Lama are different - they have a clutch that allows the engine to idle without turning the rotor).

This may not seem like much power, and for that part of the descent where the rotor is split from the power turbine, perhaps not of much consequence, but it is enough.

When the power turbine RPM is below the rotor RPM (i.e. the needles are "split", the engine is not providing any power to drive the rotor.

Most of us only care that there is a split, but don't notice that the power turbine is still turning at a fairly high speed. If the rotor RPM is reduced (as in a maximum range autorotation), the power turbine speed may be high enough to contribute power to turn the rotor.

And when the rotor RPM is reduced by raising the collective to cushion the touchdown, the power turbine will certainly contribute.

Now comes the math - the rate of descent of a helicopter in autorotation is going to depend on the weight, density altitude and airspeed. If we keep everything constant we can determine the change in rate of descent if the engine isn't producing any power. Rate of descent can be calculated as:.

$$Rate\ of\ Descent = \frac{33,000 \times Change\ in\ Horsepower}{weight}$$

Let's say we weigh 3,300 lb. (for ease of calculation), and we normally require about 160 horsepower (hp) in level flight at 60 KIAS. If the engine is producing 20 hp at idle, our rate of descent in the practice autorotation would be 160-20 =140 x 10 = 1,400 feet per minute. If the engine were really stopped, we'd be doing 1,600 feet per minute, or about 15% more rate of descent. A steeper rate of descent. Given that helicopter glide angles are already pretty steep, making it steeper at the wrong time isn't much fun.

For those who say - but the rotor is split from the power turbine and is being driven solely by the air passing up through it - why is the rate of descent more if the engine is really failed? The power turbine is still producing power that is going to the transmission - and the transmission takes power to turn things like the hydraulic pump. So, even if the rotor is split from power turbine, the power turbine is still overcoming some of the losses. With the engine stopped, that power has to come from the rotor.

But wait - there's more.

You now have to stop the increased rate of descent / steeper glide angle. This takes a bit more energy to stop than it did with the engine on, because the change in flight path is slightly greater. You'll feel like you're falling through a part of the maneuver that was straightforward in training. If you get that right, then the touchdown is another unpleasant surprise.

The rate of rotor RPM decay at the end of the autorotation will be much higher because you don't have that idling engine which can keep the rotor turning at least at idle. So, you may land a bit more heavily (harder) than you did in practice at the same weight.

The astute of you will now be wondering - how does this affect the H-V curve? Shouldn't the H-V curve be bigger than you'd suspect because of this?

In reality, the H-V curve is a very complex subject (covered in Chapter 14 through Chapter 16) and the authorities will make due notice of the differences between engine at idle and engine really off - every H-V demonstration for certification that I've seen resulted in no damage to the landing gear, let alone the helicopter, which means there was some reserve energy left in the rotor, as nothing was damaged. It would be reasonable to expect that in a real engine failure the landing gear might have to soak up more energy.

So, remember, you might be an ace when it comes to practice autorotations, but be prepared if you have a real engine failure that things might not be exactly as you had thought.

An Afterword

Note- after this article originally appeared in a magazine, one of the major helicopter manufacturers took exception[1] to the contents, and had an article printed in reply in the magazine. Their training pilots actually did comparison autorotations that showed there was no difference between "engine at idle" and "engine off" rate of descent and landings. This is to be commended, as it was a gusty move and showed to both the instructors and their students that there was no difference. The company also required their instructors to carry out one engine shut-off autorotation every year. Again, a move that is to be commended.

But the article was based on actual observations from another manufacturer's helicopters and with words in the FM that stressed the difference between autorotations at idle and real autorotations. That company's test pilot had noticed the difference and decided to test it.

Summary of Chapter 12

This chapter has stressed some of the differences that may (and I must stress may) occur between practice autorotations with the throttle at idle and a real engine failure with the engine actually shut down.

Questions

1. In a turbine helicopter will practice and real autorotations (with an engine failed) always have the same performance? (Answer will depend on what type you may be familiar with!!)
2. If you are unlucky enough to have a real engine failure but lucky enough to survive failure and see a major difference in performance, what would you do?

---

[1] in fact, pretty great exception

Chapter 13 The Height-Velocity (H-V) Curve

No discussion about autorotations would be complete without something about the H-V curve. This section deals only with single engine helicopters.

A typical H-V curve is shown in Figure 13-1 Typical H-V Curve..

Autorotational Height-Velocity Plot

CH136 Kiowa

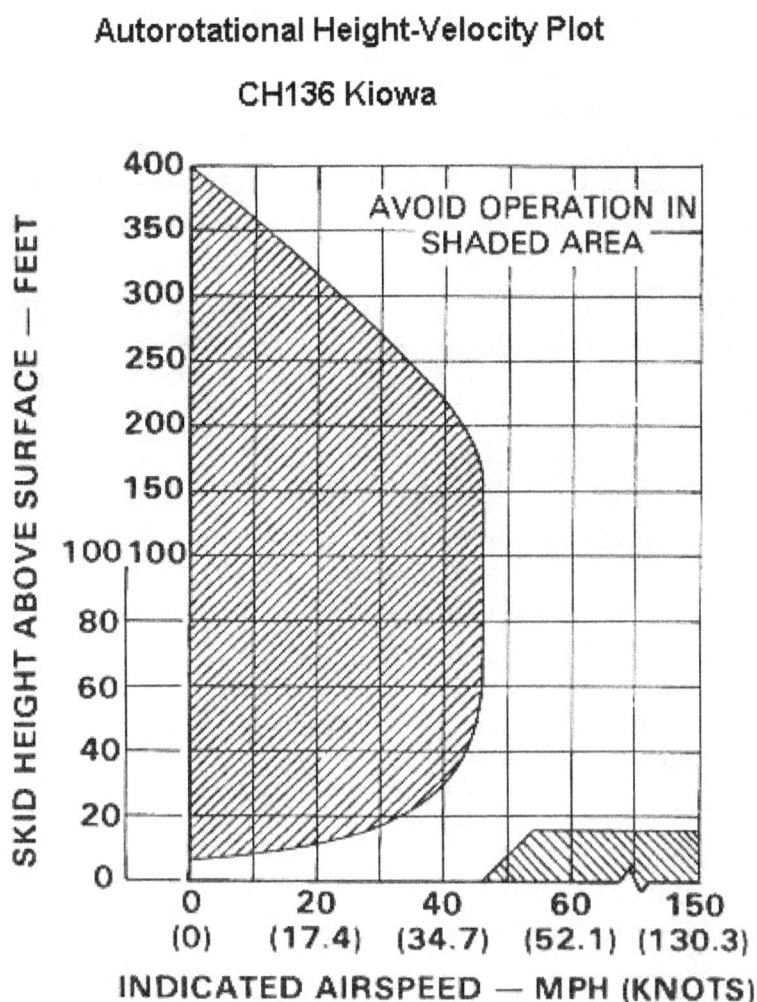

Figure 13-1 Typical H-V Curve.

The height velocity (H-V) curve has been widely misunderstood and badly written about in many magazine articles and books. I have had the pleasure of teaching experienced helicopter pilots, as students at test pilots" schools, about this curve, and would like to pass on some observations.

A typical definition of the is

-the area of heights and airspeeds within which it is difficult or impossible to safely land the helicopter following an engine failure".

67

The H-V curve, like any other performance chart, needs to have the conditions relevant to it shown with the chart. Sadly, most Flight Manuals (FM) do not show these facts. What is often not mentioned in the definition are the weight, wind, density altitude, pilot intervention time, and landing surface. The curve may not be exactly what you think it is, yet it is remarkably accurate and repeatable.

For certification purposes, the H-V curve must be demonstrated at the maximum weight for the altitude being sought for certification.

When you look in the manual, you can be pretty well guaranteed the curve is for the maximum weight of the helicopter, at a high density altitude.

The overwhelming response of experienced helicopter pilots when they had finished a demonstration of the H-V curve and how it is developed has been: "Why haven't I seen this before? Why has no-one bothered to show me what the characteristics of this helicopter were like from an OGE hover high above the ground?" Wish I knew, especially as it's a part of the flight envelope where the helicopter does unique work.

Most pilots undergoing helicopter training do not see a wide variety of autorotation entry conditions. Most are of the "aligned--with--landing--direction,-know--we--can--make--it, lower--the--collective--roll--off--the--throttle" variety. Those entries also probably have the pilot very ready for the situation.

The result is often an overly confident pilot about what happens to the helicopter from say, a 500' AGL hover when the engine goes on a long lunch break.

I know pilots who literally "beat" the H-V curve - they had an engine failure inside the avoid area on the chart. How is it possible they not only survived, but did no damage to the helicopter? When dissecting their incidents, it was apparent they were well trained in autorotative landings, knew the symptoms of an engine failure, were spring loaded to the engine failure reaction position, and had a suitable area in front of them. They reacted immediately and instinctively, knew where they were going to go and what they were going to do before the engine failed. They were also very lucky.

The H-V curve is not a curve to be treated lightly.

Be prepared:

Know what the reaction of the helicopter is to the engine failure in different airspeeds and flight conditions. The reaction of the airframe to an engine failure in a 1000' hover is very different from that at 60 KIAS in level (or descending) flight. Know what you are going to do instinctively. Know where you are going to attempt to land. Be prepared for the engine to quit. (I know of one pilot who says he's always surprised at the end of a flight if the engine didn't quit!)

If you can react more quickly than the intervention time used in developing the curve, then you are that little bit better off.

One of the better pilots I had the privilege to train was a veteran TV news pilot in a very large city. When I asked him what he considered the keys to survival, he responded -- "I never go below 300' AGL or below 50 knots. If I go below either 500' or 70 knots, I'm talking to the cameraman as we orbit all the time about where we'd go if the engine failed." Wise words.

If your helicopter flying regularly calls for you to operate in or near the H-V area - such as slinging, military missions, photographic missions and so on, then ask yourself if you know what the symptoms of an engine failure really are at all the airspeeds you fly - including the high hover. You might be more than slightly surprised to notice how your helicopter reacts when you lower the collective from 10 or 15 knots of airspeed at 1,000' AGL.

The H-V curve, by itself is not definitive in all respects of ensuring a safe landing if you're neither outside the H-V Curve, nor of certain injury or damage if you're inside it. Be warned.

Ignoring A Part of Most H-V Curves

Those of you who have compared the curve shown in Figure 13-1 Typical H-V Curve. to more modern curves will notice that the newer helicopter H-V curves are missing a portion shown above, namely the high speed, low altitude part. Why has it been left out?

The reason is that the way most curves are determined, namely using some intervention time, is, in my opinion, inappropriate for this part of the curve. I don't know of any helicopter pilots who would fly at this low height above ground at the speeds shown without being very actively involved in controlling the helicopter. Even if the engine fails, they are going to be extremely interested in keeping the helicopter from diving into terra firma, and might not even be aware immediately that the engine had failed. There is enough kinetic energy in the helicopter to control the machine, and assuming a suitable area, it should always be possible to make a safe landing.

To be perfectly correct, if you were to wait one second before carrying out any actions following an engine failure in this flight regime, you would be first, have to be asleep to not notice something wrong with the flight path, and secondly be deep in trouble. It is almost certain that any pilot would not be actively involved in flying the helicopter in these conditions and react instantly. If they weren't awake, they might deserve what happens to them.

So, for that reason, most modern helicopters no longer feature this portion of the H-V curve.

Difference between Civil and Military H-V Curves for Same Model Helicopter

Another important point for those comparing this curve to the civilian Bell 206B series (which is much smaller in size, with a lower high hover point) is that the curve shown above used a 2 second intervention between engine failure and pilot reaction on all the flight controls, where the civil version only uses a 1 second intervention on the collective lever only. A world of difference.

So What's Missing about the H-V Curve?

Aside from knowledge of what the curve means in general, the following are things I think are missing.

- The first thing missing from the FM description of the H-V curve is the relevant performance conditions.
- Second, it is developed in very little, if any wind. A problem is that the wind at the high hover points is normally unknown. In my experience (and from what I have read), the presence or absence of wind at the high hover points has not made any difference to the resulting size of the curve. Some detailed engineering comparison tests shows the high hover point can be almost independent of weight, density altitude or even surface wind -others show the same variables have a large effect on the size and shape of the curve.

- Third, the curve involves a degree of delay between the engine "failure" and the pilot taking any action. This delay is to simulate the typical pilot being caught unawares when the engine fails. Various parts of the curve have different delays, to take into account different levels of awareness of the pilot. For example, the high hover points (above the knee) have at least a one--second delay between the engine "failure" and the pilot taking any action. This is supposed to account for a minimum skill level pilot's relative ability, although many would argue it's not valid[1]
- Fourth - the landing criterion is next on the list of unlisted items. Nothing says the touchdowns must be to zero--groundspeed. Most test points feature a running landing. Don't attempt to duplicate an H-V test point and try for a zero--groundspeed touchdown. Most H-V points in certification testing have quite long ground runs.
- The fifth and final thing that's missing is training of pilots to know how to handle engine failures when operating close to, around or (shudder) inside the curve.

The obvious question is "Why isn't this information presented somewhere, like the Flight Manual?". The short answer is that the manufacturers only put in what is needed to save time, money and reduce their exposure to liability.

Development of the H-V Curve

The way the H-V curve is developed is beyond the scope of this book, but it should be mentioned that the whole process is approached with a sobering degree of caution and build up. There is a real potential to damage the helicopter as well as the crew. I do not wish to slam other textbooks, but more than one has gotten the development of the H-V curve very wrong. Did you know there have been two helicopters developed that had no H-V curve? Do you think it likely they stayed airborne after the engine failed?

One book has claimed that the curves are deliberately made smaller for marketing purposes, however I have never seen any that were unrealistic for the conditions tested. I know of one civilian machine adopted for military use that had the high hover point raised by the military when they discovered that their supposedly superior military instructor pilots couldn't handle an engine failure at the civilian-developed high hover point. I don't know what that exercise proved, except it certainly didn't result in any great improvement in safety for the military, or degradation in safety for the civilians.

The test points are approached incrementally, and more than one test pilot is involved, just to make sure the results are repeatable. Having demonstrated how its H-V curve is developed many times, I can say that for the Bell 206B, the curve in the civilian FM is very accurate.

For example, an engine failure (simulated) in the zero--groundspeed hover at 420 feet above ground can be handled by someone who knows what they are doing. However, from a entry in hover at 350' AGL, even very experienced helicopter pilots abort the demonstration point, add power and go around. Similarly 45 knots in level flight at 200 feet is OK, if you knew what to do, but 40 knots at the same height is not - the helicopter literally falls out of the air.

Legal Requirement for the H-V Curve

The legal requirement for the H-V curve comes from 14 CFR Part 27, paragraph 27.79 - but there are other paragraphs that are indirectly affected by this one. And the advisory circular

---

[1] Military criteria call for two seconds - a very loooong time indeed.

for Part 27 - Advisory Circular 27-1B has some pretty specific advice and guidance about how this should be tested to demonstrate compliance with the regulation.

It's also 900 pages long, and not the most riveting reading in the world.

The mildly interesting aspect to how this part of the H-V curve is tested is that it assumes the pilot is actively flying the helicopter and will react immediately to any thing that affects the flight path of the machine. In other words, the pilot's hands are on the controls and the pilot is properly trained to recognize that that an engine failure has occurred.

The testing also assumes a smooth, flat, hard surface below and in front of the helicopter. As if that always exists....

Part 29 H-V Curve is a Limitation

It is very important to note that Part 29 helicopters (i.e maximum weight more than 7,000 lbs), with more 10 or more passengers have the H-V curve in the Limitations section of the flight manual, and you must not operate there. Conversely, if a Part 29 helicopter has 9 or fewer passenger seats (for example, an EMS configuration), the H-V curve is in the Performance section.

Conditions for Testing

The test conditions also assume that the helicopter is at maximum weight, and worst center of gravity (typically a forward CG, as that is going to take the largest pitch attitude change to make the helicopter stop).

A maximum nose down attitude of 20° nose down is used, as the authorities consider that most pilots wouldn't use more than that steep a nose-down attitude. Remember we are not making this curve for test pilots, who've seen some pretty dramatic attitudes.

For some of the tests takeoff power is used, and it is assumed for takeoff that the pilot is hands on the controls. This means in the takeoff profile, there is no delay between simulating the engine failure and the pilot's reaction to the failure.

As with many things, there are sections of the certification requirements that indirectly impose themselves on other areas of how we should operate. Paragraph 27.1587 for example, says that any information pertinent to the takeoff procedure should be published - a bit of information more honored in the breach than the observance, (as Shakespeare would say). Paragraph 27.51 says that the takeoff path must not require exceptional piloting skill or favorable conditions and (more importantly) must be made so that a landing can be made safely at any point along the flight path if an engine fails....

Miscellaneous Points About the H-V Curve

There has been some discussion about pilots being legally charged for flying within the H-V curve and causing danger to persons on the ground. I would look at this the other way - just because there is an H-V curve, doesn't mean the people on the ground would be endangered.

What if there were no H-V curve and the engine failed? Would the helicopter not land? What if the helicopter were outside the H-V curve and the engine failed over the molten lava pits of a volcano? Would that be safe?

As another good reason to understand why you should be familiar with the H-V curve is the lesson learned by nearly every pilot for whom I have demonstrated this curve. From a high hover, when the engine fails, the first reaction of the Bell 206B is to yaw and then sink vertically. Within a very few seconds, and without the pilot doing anything to the flight controls (except to lower the collective and stop the yaw), the nose will drop to about 30° nose down, and the airspeed will rapidly accelerate towards the normal airspeed in autorotation - 60 knots is reached in very short order. The point is that if the pilot pushed the stick forward right after the engine failure on this particular type of helicopter, he might get into serious trouble - the nose would be very, very far down indeed. This is not a fault of the machine, by the way, but of the training and licensing system that doesn't require this to be shown to all helicopter pilots on this type.

Another thing to note is that if you were hovering at 100' AGL when the engine failed in this model, you would have to work to stop the nose from pitching down, and if you did let the nose drop, you would not gain any airspeed at all to start to flare in that short distance.

Another good reason to move the cyclic aft at the first sign of an engine failure.

One thing I can say about those who have seen the H-V demo - they never try hovering at higher than the low hover height without an extra-sensory awareness of the engine…

Determining a Takeoff Profile

If we return to the basic reason for determining this part of the curve, it often helps to keep the big picture clear - we want to determine a takeoff profile that will allow the pilot to be able to carry out a safe landing if there is a suitable surface underneath.

This normally boils down to a problem of too much height without enough forward speed. This means that there is not enough energy to change the helicopter's flight path to slow the rate of descent and forward speed. Better to have forward speed to be able to control the height and rate of descent and slow them both down (i.e. flaring) than to run out of energy and fall heavily to the ground.

So our test pilot has to start from a slow speed reasonably low to the ground (not much higher than the low hover height) and by increasing speed and height, find the combinations of airspeed and height which define the points above which (in height), for a set speed, would not be safe if the engine failed.

Then, the phrase used in the certification requirements (Part 27 paragraph 51) about a safe landing at any point along the takeoff path comes in to play.

The words in the advisory circular (AC 27) that shows how to demonstrate compliance with the regulations tells us that the regulators want a takeoff path that keeps the helicopter clear of this lower part of the H-V curve by a margin of 10' at any speed, and 5 knots at any height. See Figure 13-2 Typical Takeoff Profile.Figure 13-2 for an example of a recommended takeoff path.

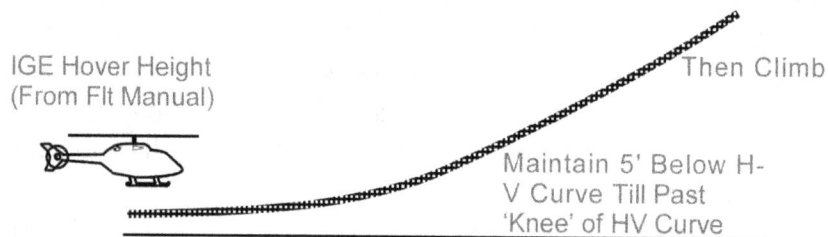

IGE Hover Height
(From Flt Manual)

Then Climb

Maintain 5' Below H-V Curve Till Past 'Knee' of HV Curve

Figure 13-2 Typical Takeoff Profile.

Deliberately Flying In the Avoid Curve

What is usually seen in normal operations is often something different than the recommended profile - for example a takeoff profile that has the helicopter climbing at a slow airspeed - passing say 30' AGL at 20 or 25 knots. While this may be done deliberately, with full knowledge of the consequences, my experience in demonstrating the H-V curve and it's development is that many pilots are unaware of the consequences of this altitude vs. airspeed profile.

Once pilots have had the H-V curve demonstrated, especially the portion below the knee, they will always use the recommended takeoff profile unless there is good reason not to. Obstacles such as trees, buildings, and so on may dictate the takeoff profile, but they will do it with a more complete knowledge of the consequences of an engine failure.

And if they are forced to fly in the H-V curve, they know to be especially alert for the symptoms of an engine failure, and to apply aft cyclic at the same time as lowering the collective. This way they can maintain rotor RPM and control all the way to when they need to raise the collective.

Helicopters with No H-V Curve

There have been at least two helicopters that showed no requirement for an H-V curve. One was a tip-jet helicopter, and the other was a very specially modified version of a production machine that had about 50 pounds of extra weight added to the main rotor blades. It didn't mean that they would stay airborne following an engine failure, merely that neither height above ground nor speed seemed to matter as far as landing safely. On suitable terrain - I'll leave you to think of the consequences if the terrain weren't suitable.

Summary of Chapter 13

It should be obvious that autorotations need a lot more attention than they receive in most training establishments. As previously stated, power recovery autorotations have 95% of the learning and 5% of the risk of full, engine--off touchdowns, especially when the judgment of airspeed and glide path angle are the really important parts of autorotations. Nearly every variation in this chapter can be carried out quite effectively to a power recovery, with little risk and lots of teaching.

There is no such thing as the "textbook" autorotation - mistakes will be made in every one. The main thing is to be able to understand the variables at your disposal, the ways you can use them and the overall aim - to walk away from the helicopter.

You should be able to identify and correct the inevitable errors quickly.

One of my old flight instructor comrades summed up a pilot's progress at autorotations nicely: "He corrected for his own mistakes immediately."

It should also be obvious that the H-V curve isn't what it always seems to be. Be prepared!

Questions

1. What conditions of weight, density altitude and wind are used to develop the H-V curve shown in the flight manual?
2. What power is used for the portion from the low hover to the knee of the curve?
3. What power is used for the portion above the knee of the curve?
4. Is the collective allowed to be lowered before being raised for the low hover point?

5. Is it likely that you would be prepared for an engine failure if it were the real thing?

6. Why should a takeoff profile stay clear of the H-V curve avoid area?

7. Where is the H-V curve in a helicopter with more than 9 passenger seats?

Chapter 14   The H-V Low Hover Point

The Low Hover Point

Engines power our rotor blades. But, being of earthly construction, they, on occasion, are bound to fail. And, while we hope the engine won't quit unexpectedly before that time is up, they sometimes unfortunately do. Where you are when the engine quits, not just in regards to the terrain below, but also the height above the ground, will make a large difference in your outcome.

How it Begins

Those of us who have any experience flying helicopters will have done engine-off landings from the hover --- it's where you begin to see the complexities of arriving safely on earth without power. If you're like me, you probably didn't know the H/V curve existed when you started this interesting adventure. Your instructor would have taught you the fundamentals of "stop the drift, stop the yaw and cushion the touchdown," or whatever their own mantra was. You learned to react and judge the events to a suitable level of proficiency, then progressed to more advanced things.

Perhaps your instructor took you to a "high" hover and showed you that when hovering this high above the ground and the engine fails, you should first lower the collective slightly and then raise it to cushion the touchdown. Confident in this newfound skill, you were (and maybe still are) happy to hover at 10 to 15 feet above the ground all the time. You felt you could handle an engine failure at nearly any height above the ground, because you'd practiced it in training.

If you're like me, and later became an instructor, you probably taught students how to do this. Hopefully, though, you're not like me, who demonstrated a hovering engine failure to a student from a significantly higher height. (It was a firm landing, in case you were wondering.)

Some of you may even have tried to demonstrate landing from an extremely high hover with an exquisitely timed application of collective. Sadly, you've been misled about what will happen in an actual engine failure in the hover. Before the reasons are explained about why the low hover point exists, let's look at why you may have developed a misplaced sense of confidence.

Misplaced Confidence

It all has to do with how you were trained. As a student pilot, you would have either been informed or in some way expected that every training trip would include at least one engine failure. The instructor probably briefed you on the maneuver immediately before either demonstrating it or making you do it. There may have even been an announcement of a practice engine failure. In short, you knew what was coming. Even better, you knew exactly when the failure was going to happen, because you had your hand on the throttle and could feel it move.

Since the only safe way for the instructor to simulate an engine failure in most training helicopters is by putting the throttle to idle, you knew what an uncommanded throttle movement meant. If only the real world could be so kind.

Aside from some noise that might indicate something is not right in the engine room, there is often little warning that the engine is going to quit. Every helicopter pilot I've talked to who had an engine failure (including me) was quite surprised to no longer have power.

When an engine fails for real, it normally takes a few seconds to react to this new situation. After the "this-can't-be-happening-to-me-now" thought, instinct takes over and you have to deal with the cards you've been dealt.

Think about this for a moment. You won't know when the engine quits and yet you've been trained with the assumption that you will know.

Any questions?

Fortunately, the authorities have recognized this in the development of the H/V curve, and account for the surprise of a real failure.

The low hover point is a perfect example of how they try to protect pilots from themselves.

Defining the Low Hover Point

The first assumption used when developing the H-V curve is that the pilot will be hands on the controls, and will react to the situation with little or no intervention time. The statements in the Advisory Circulars that tell how to test this state "no intervention time" is needed between engine failure and pilot reaction.

Next comes limitations for the test method --- the helicopter needs to be at maximum weight. For most training helicopters, having two people on board and full fuel puts it almost at maximum weight, so our training method probably comes close to meeting this requirement.

But, carrying this out at sea level is different from operating at high density altitude The requirement is to conduct the tests at 7,000' density altitude. Not much training gets done at these conditions, which is perhaps fortunate.

Our test helicopter is now set. What exactly is it we're trying to find again? Right, we're trying to find a height below which we can survive the engine failing.

We start from a very low hover, fail the engine and see how the helicopter handles. From a very low hover, we simply pull up on the collective to cushion the landing. There's no point in lowering it, even if we wanted to, there is no need and no benefit. Satisfied with that height, we increase the height above ground and repeat the test, then repeat it again at higher heights. The loads on the landing gear are monitored pretty carefully, as is the rate of descent and a ton of other parameters.

At some point, the thought crosses the mind of the test pilot that lowering the collective following an engine failure might be feasible, but the advice on how to test says the collective can't be lowered first. Why no lowering of the collective? Because all the evidence from real engine failures says there isn't enough time to do this - by the time the pilot recognizes that an engine failure has happened, there isn't enough rotor RPM or time to lower the collective and then raise it again. All you can do is try to arrest the rate of descent with an up collective movement.

That's why nearly all helicopters have a low hover point of between six and eight feet above ground. Above that height, if all you can do is raise the collective to cushion the touchdown, you will probably not have enough inertia or lift in the blades to be able to arrest the rate of descent sufficiently to prevent damage to some part of the helicopter.

Applying This to the Real World

Having demonstrated the H-V curve build up quite a few times, I can tell you that the low hover point is quite sharply defined. One height looks fine, the next increment looks a little more dramatic and the end point is usually defined as: "I'm not going to hover any higher because I don't think I want to know the result if the engine fails!" I know of no other way to describe it than "the bottom just falls out" from that height.

My experience in training and demonstrating the H-V curve is that weight and density altitude don't appear to make a significant difference to this height. Regardless of how light the helicopter is, or how low the density altitude, when all you can do is raise the collective to stop the rate of descent, you don't want to hover much above this height.

So, if you're proud of your ability to do hover engine failures from a height above that of the published low hover height in H-V curve, remember that the real thing isn't the same as how you've been trained or how you might practice it.

Why the practical test standard does not call for hovering engine failures[1] to be demonstrated at the low hover H-V height, with only an upward application of the collective is another mystery..

If you see someone hovering at 15 to 20 feet above ground level, you might just cringe as you watch them, secure in the knowledge that they know not what they're setting themselves up for.

In the next chapter, we look at that space between the low hover point and the knee of the curve --- an adventure in takeoff technique!

Summary of Chapter 14

The low hover point of the H-V curve is defined by the ability of the landing gear to absorb the rate of descent more than pilot skill or the rotor's ability to stop the rate of descent.

Questions

1. Is an engine failure in the hover an autorotation?
2. What weight and density altitude conditions is the low hover engine failure height in the flight manual demonstrated at?
3. Does Density Altitude and weight make much difference in the height from which it's safe to hover in the event of an engine failure?
4. In demonstrating the low hover point for certification, is the test pilot permitted to lower the collective prior to raising it to cushion the touchdown?
5. What is the rationale for the technique described in the Advisory Circulars for this point as far as collective movements are concerned?

---

[1] Oops -- nearly said "hovering autorotations"!!!

Chapter 15 Takeoff Portion of H-V Curve

Having taken the low hover point, how it is developed and what it means to the pilot and beaten it to a pulp in the last chapter, it's time to move to the next section of the H-V curve - namely that part from the low hover up to the "knee" of the curve.

This section discusses mildly interesting aspects of testing, and some very interesting observations on how we should operate our helicopters.

As in the last chapter, let's go to training as our first step. As before, the mere fact that the flight is a training trip for most of us means we're already prepared for an engine failure. The first twitch of the instructor's hand to retarding the throttle means we can now unleash those prepared reactions to demonstrate our superior flying abilities.

Ah, If only the gods were so kind to us when a real engine failure occurs. Experience shows that, aside from those gifted with ESP or those who listen to the still, small voice or have some other indication of something going wrong, we are all quite unprepared for the real engine failure. So perhaps using a "normal reaction time" (i.e. no time between failure and reaction) is a bit optimistic.

On the other hand, I don't know too many helicopter pilots whose hands are not on the collective and cyclic during takeoff....

But the regulations being the way the game is played, we use those rules. For this part of the H-V curve, the power to be used is takeoff power (or some other power the manufacturer designates, such as hover power plus 10%), the weight is maximum weight and the terrain is a smooth hard surface suitable for landing. The test pilot attempts to attain the height, airspeed and power setting all at the time and then simulate the engine failure and carry out a landing.

A busy time in the cockpit, as I can testify. Airspeed indicators are not particularly repeatable in this phase of flight, heights above ground have to be estimated as there is no production equipment that is really accurate enough to be used (there will be flight test equipment that is accurate, but it is for recording data), and it all happens pretty quickly.

Engine Failures Along This Section of the H-V Curve

For certification, takeoff power (or some other value, such as 10% above hover power) is used. This means high collective angles, higher than would be used in most normal operations. The rotor decay following the engine failure is higher than normal, however that appears to be offset by the higher inertia than might be seen in normal operations.

That's confusing, so let's explain.

The higher than normal power setting means the acceleration is going to be more rapid than normal. In this low speed environment, the Pitot static system is not very accurate in measuring airspeed, so a more rapid acceleration means the airspeed indicator is going to be slower to respond than normal. This means that when the airspeed indicator reads 30 knots, for example, the helicopter is actually going to be going faster than that -- let's say the actual airspeed is 40 knots.

In a "normal" takeoff, when the airspeed indicator reads 30 knots, the helicopter may only be at 35 knots as it's accelerating more slowly than the test helicopter would be.

With the higher real airspeed, the flare effectiveness of the helicopter will be greater than the "normal" helicopter. However, the "normal" helicopter may not be using takeoff power and so will have a lower rotor RPM decay rate. It all seems to even out.

But the point for the operational helicopter pilot is that the reaction to the power failure needs to be instinctive and immediate.

What Does this Mean to the Pilot?

The H-V curve doesn't exist in isolation. It has a two-fold purpose.

The obvious first reason is to delineate the combinations of (air)speed and height that are (within the conditions of test) unsafe to be in if the engine fails. The not-so-obvious second reason is to provide some guidance for a takeoff corridor that is safe. That's the main purpose of this lower boundary to the curve.

If there is open space that permits an acceleration and climb-out keeping clear of the H-V curve, then, that's the prudent way to get to forward flight.

Surviving in the Real World

And what should you do in this area if you're forced to takeoff with more altitude than airspeed?

If you're faced with tall trees on your takeoff path, your choices are obviously dictated by those obstacles. But, if you have a clear area in front of you, a takeoff profile that minimizes your exposure to danger is prudent.

The H-V curve takeoff portion is flown using takeoff power, and unless you're heavily loaded, you don't need to use that much power. But you do need to be aware of what you need to do to handle an engine failure.

Obviously no advice will cover every situation. The keys to survival in this case have to do with the obstacles and ground in front of you. "Better to run into the far wall at a small forward speed than the close one at high speed" was advice from my fixed wing days that still works. For a helicopter, since we can control our forward speed without danger of falling out of the sky in a stall, change that to "maintain the rotor RPM for control, and make the vertical speed and forward speed as slow as possible prior to hitting something".

Without rotor RPM you have no control over your rapidly looming destiny.

There may be little or no flare effectiveness at airspeeds below the knee of the curve, so the collective is going to be the main (only?) method of controlling the flight path. A slight nose up attitude may help make the collective application more effective at slowing any forward speed. If it's a smooth surface below, the landing gear may be able to soak up a lot of the vertical speed, but that means getting the flight path direction and the heading of the helicopter to be the same so the landing gear can work properly.

Even if you have a clear area ahead that will let you use the recommended takeoff profile, the ground may be less than inviting - anything other than a smooth, hard surface (and there's lots of that around) means you would probably like to touchdown with no forward motion - and that's not always guaranteed even if you are following the recommended profile.

Lots of gotcha things here - while it's not perfect, the recommended profile is a good place to start, and the successful pilot will be the one who knows what to do when faced with situations outside the "normal" (whatever that is!).

Summary of Chapter 15

The takeoff portion of the H-V curve is a busy time for the test pilot but makes life easier for the rest of us by providing a safe takeoff corridor when conditions permit.

Questions

1. What power level is used for the takeoff portion of the H-V curve?
2. Why is the "normal pilot intervention" time in this portion of the curve accepted to be no intervention time?

Chapter 16   The H-V Curve High Hover Point

In this chapter on the H-V curve, we're going to work on the most interesting part - the high hover area.

This part is the most interesting as it adds the dimension of height.

With height comes potential energy - another variable to deal with.

Remember, the H-V curve defines the combinations of height and speed from which it is not possible to make a safe landing following an engine failure. So far we've looked at the low hover and the "takeoff" portion.

We're going to start at a height above the high hover point, at cruise airspeed and look at the energies available. Then we'll reduce speed to the hover to the same height. For ease of argument, we'll say that our cruise speed is 100 knots and the high hover point on the H-V curve is 500' AGL. Like all mathematicians, we'll also start with a mass for the helicopter of 1 (assuming nothing falls off the helicopter, the mass doesn't matter for this discussion). We'll also assume that we have an "ideal" airspeed in a power off descent of 60 knots.

Energies to Play With

If we start from this cruise condition we can work out how much energy we have to play with. There are three types of energy we can use --

- Potential energy, the energy due to height above the ground,
- Kinetic energy - the energy of speed with respect to the ground, and
- Rotational energy- the energy stored in the rotor.

The formulae for these different conditions were given earlier.

To simplify things, we'll assume that the rotor RPM is constant throughout this maneuver.

In our cruise condition, assuming we react (more or less) immediately to the engine failure and don't lose any rotor RPM we can trade speed (Kinetic Energy) to maintain height for a short time while we decelerate to the "ideal" airspeed. We start at 100 knots with (100^2) = 10,000 units of Kinetic Energy, and 16,000 (500 x 32) units of potential energy, for a total of 26,000 units. After that, we trade Potential Energy for airspeed and rotor RPM, until finally for a flare height of 50' AGL, at 60 knots we need to have 3,600 (60^2) Kinetic Energy + 1600 (50x 32) Potential Energy for a total 5,200 units (we'll leave the rest of the flare and landing out of this discussion).

Now let's consider the other situation - a hover at the same height of 500' AGL. In this case we'll talk about a zero-groundspeed hover for those performance purists. What energies do we have to use? Obviously, only Potential Energy (16,000 units) and Rotational energy! We must maintain rotational energy in order to have any hope of control (and survival), so we're left with only Kinetic Energy.

So from a zero speed[1] hover at 500', we have to go from a total energy available value of 16,000 to a flare height value of 5,200 in order to safely land.

Seems like we could arrive at this energy value from a much lower height than 500' AGL.

---

[1] airspeed or groundspeed, it doesn't much matter, but I always use groundspeed

Unfortunately, this simplistic approach ignores a lot of things. We're not going to try to develop the mathematical equations to show this.

The reality is that it takes a lot of energy to accelerate and overcome the drag of the rotor and airframe, and for our example helicopter, the high hover height was determined to be around 450' AGL.

How was this derived?

By starting at a much higher hover height, and seeing how much height was lost to re-accelerate the helicopter to 60 knots. The initial height for starting this exploration was around 1,000' AGL. It was found, with a one-second delay between simulating the engine failure and lowering the collective that it took about 300 feet to get the rotor RPM back to the "normal" range and the airspeed to 60 knots.

After several cautious reductions in hover height (to take into account slight wind shifts and ensure it wasn't just luck), the lowest height at which the test pilot was willing to hover and attempt a landing was 450'.

Along the way an interesting thing happened...

First -- the maximum nose down attitude the Advisory Circular recommends is 20° nose down. The first time you see this in a helicopter, it's a bit unusual, but you can get used to it pretty quickly. But the test is for the pilot who's never seen this before, not the seasoned test pilot. Even limiting the nose down attitude to 20° nose down results in an initial rate of descent of about 3,000 feet per minute. Much faster than the "normal" rate of descent of 1,600 feet per minute.

Second -- the slow response of the airspeed system was mentioned in the takeoff part of the H-V curve. It's even slower to respond here with an added reason that there is just not much dynamic pressure to measure.

[50 Knots Indicated Airspeed is Really 60 Knots]{50 Knots Indicated Airspeed is Really 60 Knots.

The "normal" airspeed we want to be able to flare effectively is 60 knots, and we certainly don't want to be too slow before starting the flare. But even with the very nose low attitude and the sensation that we're doing at least 60 knots, the airspeed indicator only reads 50 knots. Something in your other senses says this slower indicated airspeed is still OK for stopping the flare, which works as nicely as 60 knots in a normal autorotation. But why only 50 knots when everything else says it's 60? Remember that an airspeed indicator only tell the difference between static and dynamic pressure. In this case, the air pressure is changing very quickly (3,000 feet per minute is about 34 miles per hour), and the static part of the airspeed system can't keep up. You really are doing at least 60 knots (I checked it with a GPS).

Back to Determining the High Hover Point

So, we start from our high hover, simulate the engine failure and wait for one second and briskly lower the collective. Now what??? This is where training and experience are essential.

Having done over 100 demonstrations of the H-V curve to students, and several civil certifications of the H-V curve on a variety of helicopters, the reaction of the helicopter in the next few seconds is most illuminating! Some machines drop vertically for a short eternity and then the nose drops all on it's own. And continues to drop to what at first is an alarming angle, but the airspeed will rapidly increase with this nose-low attitude.

And here is the reason for having training and experience - if you didn't know the nose was going to drop all on it's own, some forward cyclic would probably be added to get things going the right way - and this is not something you want to do, because this control input would make the nose go a lot farther down than you might like.

"Way Behind the Machine

Then there's the reason called "comfort level". The first couple of times most pilots see the reaction of the helicopter to an engine failure in the high hover they are, quite frankly, way, way behind what's going on. A lot of things happen in a very short time, and they can't see or remember most of what's happened. By the third or fourth entry (remember these are demonstrations in developing the H-V curve, so we start at about 1,000' AGL and work our way down), they are:

a) more relaxed and

b) much more aware of what's going on.

Instead of facing the situation with fear and trepidation, they can now face it with knowledge and awareness of what needed to be done.

And one more thing comes from the demonstration - respect for the situation.

Continuous Maneuver

An engine failure at the height that defines the high hover point is a continuous maneuver - entry, nose drop, helicopter accelerates and reaches the speed necessary for flaring at the same time the flare begins. An exciting ride that takes just about as much time as it takes to read this paragraph.

The final thing about the high hover part of the H-V curve that I want to mention is that there isn't a lot of margin for error. If you are lower than the high hover point and don't react immediately, you will probably not have a successful landing. Demonstrating an entry from 350' (about 50' below the high hover point that was OK) was always instructional. Even inexperienced, non-pilot engineers would say "go around" within seconds of the nose going down when being shown an entry from a height not far below the flight manual high hover height. It was that obvious. Fortunately the engine was still available and we could go around.

An Important Point to Remember

The landing surface used to define the H-V curve is well defined in the Advisory Circular- it must be smooth, flat and hard (pavement or concrete is preferred). If the foregoing points weren't sufficient to convince you that a smooth, hard surface is essential for training for autorotations, then the Advisory Circulars on certification of helicopters (AC27-x and AC29-x) might get your attention. But what it means is that there is no requirement for the landing to be to a zero-speed touchdown. Most H-V demonstration points for certification end up with running landings. So, if you're hovering over relatively inhospitable terrain, and feel the need to be able to land with no forward speed, it will be necessary to add a fudge factor to the high hover point.

Summary of Chapter 16

The high hover point of the H-V curve is a point worth respecting.

Hovering at a lower height above ground pretty much dictates that an engine failure, even over suitable terrain, will not be a pleasant experience.

Questions

1. Is the high hover point a good place to hover?
2. Is it a good idea to spend much time hovering at heights between the high and low hover points?
3. What happens if you are hovering below the high hover point's height above ground, and the engine fails?
4. What size of H-V curve do you use if you are operating at a density altitude above the (unstated) height used for the published H-V curve?

Chapter 17 Different Types of Autorotations

There are undoubtedly many types of autorotations. Slightly less common but useful ones I am aware of are the "constant attitude" and zero/low speed autorotations. One is particularly good if you are flying at night or in cloud in your single engine helicopter or over water in your float equipped helicopter, and the other may be just what you need to get to somewhere directly below you or into a very small hole in the jungle.

Constant Attitude Autorotations

What to do if the need arises to land without assistance from the engine when you are not sure of the ground underneath, and have no choice about the landing site? For example, you're over some pretty swampy land that wouldn't let you do a run-on landing.

The answer is in a little known technique called the "constant attitude autorotation", invented (I think) in the UK. The name derives from the constant attitude maintained throughout the maneuver, although the name "constant airspeed" might just as well be used - there is no flare at the bottom.

The technique is simplicity itself- turn into the surface wind, reduce airspeed to some low value (35 to 40 KIAS, plus about one-half the wind speed, and wait for the ground to rush up (this will require lights at night..). When the ground rush is quite apparent, raise the collective. There is not enough energy to flare, and the slight aft tilt of the rotor and fuselage will decelerate the aircraft quite a bit. The combined effect is a landing with a minimum of forward and vertical speed - a mostly survivable situation.

If you wish to err, do so to a slightly higher airspeed. Decreasing airspeed to less than 35 knots quickly increases the rate of descent and makes directional control more difficult. It is also difficult to hold slower airspeeds due to errors in the pitot static system.

If the airspeed is permitted to decrease too much, regaining any airspeed takes a large change in pitch attitude and considerable time and height -- the pitot system is very, very slow to respond.

The effect for most of the descent is quite serene - there is little sensation of plummeting towards the earth. At about 100' AGL, this starts to change dramatically - the ground rush become apparent in a real hurry. The philosophy of the constant attitude autorotation is that at night with the landing light on or in IMC (which could be clouds all the way down to low altitude), the ground would start to make its presence known at about this time / height.

In daylight (the best place to practice this, aside from a simulator), wait until the ground rush is really quite pronounced (at about 50' AGL) before starting to feed in collective, slowly at first, and then as rapidly as needed. The amount of collective to apply is quite natural, but there is little room for error.

It should be quite obvious to even the most casual observer that there is no point in flaring. From a technical point of view, there is not enough kinetic energy to change the flight path and it would only change the attitude on ground impact.

Do not flare, accept the slight forward speed!

The logic for this maneuver is as follows: at night or in IMC or over a large body of water, there are not enough cues for the pilot to judge height above ground for a flare to stop the rate of descent.

A misjudged flare might mean hitting the ground with high forward speed as well as a vertical speed, or flaring too high and killing all the forward speed at too high a height. Either is worse than hitting the ground with very little forward and relatively low vertical speed.

So the constant attitude autorotation is a compromise.

The constant attitude autorotation also is very easy to teach, and provides a pretty reasonable chance of survival. It works especially well if you're flying a fixed float equipped helicopter over a large lake with not many visual references to help judge height above the lake surface. Pull up on the collective when you can see the very small wavelets on the water!

For those who say the constant attitude autorotation is silly - who would want to fly a single engine helicopter in cloud or at night? Not only do people fly in clouds and at night in single engine helicopters (it's normal in some militaries), but multi-engine helicopters can also have emergencies where an autorotation is the only way out.

And it's a very useful tool to have in case you ever need it.

Zero / Low Speed Autorotations

Two variations quickly demonstrate the range of possible methods to arrive at a chosen spot. These are the zero--airspeed (or low airspeed) and the maximum range autorotation. Since maximum range has been flogged mercilessly later, a quick word now about the zero--speed autorotation is in order.

The keen-eyed amongst you will have noticed by now that I have slipped from my normally boring preciseness about saying zero-airspeed or zero-groundspeed. There is a very good reason for this.

The first problem with this maneuver is that we can't measure zero airspeed ("Pitot Systems" in Cyclic and Collective will convince you if don't already know the reasons) The good news is that zero--groundspeed is close enough for the purposes of demonstration. A second equally important point is that this maneuver is not kept to zero speed all the way to the ground. The maneuver is actually a three-part sequence -

- deceleration to zero (or very low) groundspeed,
- descent at zero (or very low) groundspeed, and then
- re-acceleration to "normal" airspeed.

I can't think of a short, snappy name for a three part maneuver and "zero / low speed" defines it well enough.

This concept is best demonstrated by an entry from a higher--than--normal height AGL to give time and height to show the effect - I find 1,500' AGL overhead the entry point for a "normal" (i.e. 500' AGL straight-in) autorotation to be suitable. Figure 17-1 shows a typical sequence.

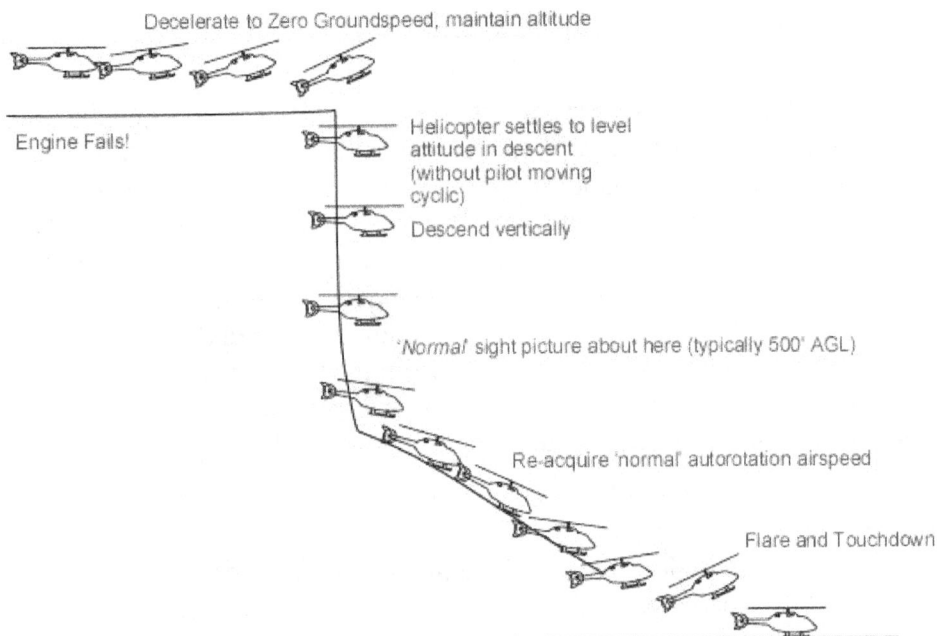

Figure 17-1 Typical Zero / Low Speed Autorotation

After the simulated engine failure (aft cyclic and down collective), decelerate the helicopter, maintaining altitude until zero (or very low) groundspeed is reached, and maintain this attitude (not altitude).

Watch out the side window for drift (forward only please - never aft).

Note groundspeed, not airspeed is used. Once the groundspeed has stabilized at a low or zero speed, don't mess with the pitch attitude. Descend in this leisurely fashion until the nominated landing spot appears to be approaching the "normal" sight picture position. Then, briskly and positively lower the nose (estimates are 15 to 20° nose down) to re-acquire a "normal" autorotation airspeed. During this acceleration, the rotor is off--loaded and the rotor RPM may decay slightly. Don't worry - the helicopter is accelerating towards the flare point with a more--nose--down attitude than normal. The pitch attitude will be more nose-down than in a "normal" autorotation. When the nose is raised to stop the rate of descent (the flare), the rotor RPM will recover, (and probably go above "normal" rotor RPM). This is due to the larger change in pitch attitude from the descent to the flare than in a normal autorotation. In fact, if the flare is really abrupt, you might consider raising the collective slightly to offset the effect of the smaller radius of the pull out - the rotor RPM will not increase as much, and somehow, it "feels" OK. After most students see this once, the reaction to raise the collective in a sharp pullout becomes instinctive.

It is important to accelerate to a reasonable airspeed in the re-conversion, at least the "normal" airspeed used - don't worry if you get too high an airspeed, but do get at least the "normal" airspeed. If you don't get enough airspeed, the flare is going to be very ineffective at stopping the rate of descent.

It is possible to descend vertically in autorotation, but don't try to take this all the way to the ground unless you are very practiced.

The only time I would consider this to be acceptable is the final part of an autorotation into a clearing in the trees. Without lots of vertical references, it is very difficult, if not impossible, to know when to raise the collective to stop the rate of descent. (I misjudged this once, and was very lucky…)

Could you do a zero airspeed descent all the way to the ground following an engine failure at a high hover? Theoretically yes, but I don't know anyone who would want to for several reasons. The first is that the rate of descent in a zero airspeed autorotation descent is going to be higher than at the "ideal" airspeed, by at least 25%, (if not more). The second reason is that trying to judge when to apply collective pitch (the only method you have to slow the rate of descent) while descending vertically is an expensively acquired skill - we humans are much better at judging height when there is forward motion than when descending vertically[1]. And the third reason is that the transition from the hover to a zero airspeed descent is a bit tricky - most helicopters have a natural tendency for the nose to drop due to the horizontal stabilizer pushing the tail up and the nose down with any purely vertical speed.

The zero / low speed Autorotation is a useful tool to help control the overall glide path to arrive at a selected landing site.

While it is quite safe to use zero / low speed to help get to a desired spot, it's normally not practical to try to land from such a vertical descent, with one possible exception.

The Hole in the Forest Scenario

If you're over a large forest of tall trees with only occasional holes, and your engine fails, it appears you've got quite a problem. The holes are probably too small to allow you to do a normal flare when you're close to the ground -- in fact, the only way you can get into them normally is from a hover with a slow descent. Hmmm. Bit of a pickle!! Do you flare into the top of the trees and hope they break your fall? Is there another solution?

If you can judge your flare at tree top level so you arrive at zero groundspeed over the top of the hole, you may be able to extricate yourself from certain doom. Once you've stopped the forward speed, in this situation you don't need to cushion your touchdown until you're close to the ground, and you still have lots of rotor RPM. Let the helicopter settle into the hole and using all the good cues provided by the trees, judge the collective application at the bottom.

And leave someone else to figure out how to get the helicopter out of the hole….

Summary of Error! Reference source not found.

This chapter has touched on two variations on how to get from the air to the ground if the engine fails when you're in cloud or at night or over the water in a float equipped helicopter.

Questions

1. Is the "normal" autorotation the only technique you can use to get down to the ground?
2. What airspeed would you use for a constant attitude autorotation in a 5 knot wind?
3. What defines when to apply collective in a constant attitude autorotation?
4. Describe how you would do an autorotation into a suburban backyard that is surrounded by fences.

---

[1] We have eyes in the front of our head, not on the sides like birds….

Chapter 18 The Problem with Training For Engine Failures

Modern engines are quite reliable, which is fortunate for safety and peace of mind, but unfortunate in another sense, as it lulls us into a sense of security that may be misplaced. Engines don't give the same warning as instructors, and a real engine failure will be a surprise.

The aim of practice is to educate the student to what should be done in real life, and make the necessary reactions as instinctive as possible.

A "real" engine failure may well be different in terms of engine deceleration time, warnings and so on, but this depends upon the type of engine.

There is an important difference between "practice" and "real" engine failures. With "practice" engine failures, the student (or pilot being checked) may have greater warning of the failure than the "real" thing.

The first reason is that the mere fact that it is a training or check flight (and includes autorotations) prompts the student to expect engine failures. Additionally, the instructor may announce the failure prior to closing the throttle, or the student may sense the instructor getting ready (telegraphing his move by a change in posture, etc.), or more commonly, the student feels the throttle move. In any case, the effect is the student is ready and waiting for the engine failure.

Oh, that the real world provided the same warnings....

It's Training - There's Going to be Engine Failures

The first thing to remember is that unless your engine gives very clear and early indications of failing, nearly everyone is surprised by the engine failure. (I know I was!) This adds the first bit of difference from what we've practiced. Every time you go on a training trip or a check ride, you know you're going to get engine failures thrown at you.

How realistic the engine failure scenarios are will depend on where you are. Some folks only ever get the "60 knots, lower the collective and roll the throttle off" type of entry. Others will get something a bit more unannounced "Engine failure - go". Some organizations will only do it at an approved area for landing. But regardless of how it's done, you are spring-loaded to expect practice engine failures.

That's the first unrealistic thing.

More Unrealistic Things About Practice Autorotations

Always Starting from a Safe Place

The second thing that's unrealistic is that unless the instructor has a death (or serious injury) wish, they'll probably be started from somewhere pretty safe - outside the H-V curve, for example, or hopefully over very hospitable terrain. If you can guarantee you'll always operate in those conditions, then this way of the training may be realistic. On the other hand, if you routinely operate within the H-V curve or over inhospitable terrain, then you need to be aware that your training and experience may not have prepared you for the real thing.

Feeling the Throttle Move

The student probably has a left hand that is welded to the throttle, especially during training trips. The student will feel any movement of the throttle[1] by the instructor, and it can only mean one thing -- practice engine failure! So the student is pre-warned by an uncommanded throttle movement.

Intervention Delay Time

How much time does the pilot have between when engine failure and when the collective must be lowered? Obviously, in single engine helicopters, not a long time, and not all helicopters are the same.

In some helicopters, the rate of rotor RPM decay following an engine failure is so rapid the pilot almost needs to have extrasensory perception (ESP) to know the engine is going to fail. The (now thankfully retired) Westland Scout operated by the British Army Air Corps comes to mind as one of these.

In some other helicopters, from a low hover the pilot could almost step outside, have a coffee and a smoke [2]%, step back inside, get seated comfortably, enquire about the family and children of the copilot, and still have lots of rotor RPM remaining to cushion the touchdown.

The same things happen in forward flight. Some helicopters have a long time between engine failure and required pilot reaction, and others have a short time. The inertia of the rotor blades has a lot to do with the time available between failure and required pilot reaction, or intervention delay time.

The intervention delay time is measured in the development and certification testing of the helicopter, and results in determining where low rotor warning lights and horns are activated.

Intervention delay time is not something the pilot must use - if the collective can be lowered at the very first sign of the failure, the pilot is in a better position than waiting until the low rotor horn begins to sound.

But the mere fact that it's training means that your intervention time on those flights will be much shorter than when you're not expecting an engine failure. I know.

Run-Down Time of the Engine

The reaction of the helicopter to a sudden power loss (for example, a driveshaft failure or engine seizure) may be different than the way engine failures are commonly simulated in training.

The reaction of two similar helicopters (the Bell 206BIII and the 206LIII) to simulated engine failures is quite different - the BIII has a leisurely wind-down of its engine (with 6 axial and one centrifugal compressor) compared to the rapid wind-down of the engine on the LIII (single centrifugal compressor). You can be assured that the certification flight-testing will have been done both ways. But rolling the throttle to idle on either is not the same as a drive shaft failure! More good reasons to learn engine failures in a simulator! See Chapter 19.

---

[1] Aside from those that have a governor (as in the R-22/44 series) would make.

[2] Back in the days when pilots smoked, of course.

Familiar Surroundings

Not only is it a training trip so you know things like engine failures are going to happen - you're also pretty familiar with the surroundings.

After a few autorotations, you know the cues that exist alongside the landing area for height, and you know what the wind is generally doing. It's difficult to not pick up those small but important items.

The place where you end up making an autorotative landing is almost certainly going to be much, much different than your training area!

Other Things That Will be Different

When a real engine failure happens, there are a number of things that are going to be different from training, aside from the surprise factor.

Weight

Training is typically carried out at a reduced weight, in order to save wear and tear on the structure. A real engine failure can happen at any weight - so lots of performance effects will not be the same.

Rotor RPM may be higher, as just one example.

Density Altitude

The density altitude where a real engine failure can happen is not dictated by the availability of a suitable landing spot for training.

Know how a higher density altitude will affect rotor RPM and collective position.

Wind

While wind and your relative position to a suitable landing spot should have been covered in the training scenarios described earlier, the engine won't wait for the wind to be "right" when it decides to fail.

Passengers

In training, it's normally (and should be) only you and the instructor.

Passengers may be alarmed by the sudden turn of events, and a bit of hysteria on their part may be present. Know how you're going to handle that before you go flying!

Summary of Chapter 18

Training is a necessary thing! It would be foolish to not train someone and then expect them to react properly when a failure occurs.

But training has to be realistic in order to be effective. Too realistic and it becomes overly dangerous. The balance between realistic training and safety is never easy, and this chapter points out some of the areas where training in autorotations is artificial.

Is it likely that a student will be more prepared for an autorotation during a training trip than when solo?

Does autorotation training that begins with the collective fully lowered represent how the helicopter will react when a real engine failure occurs?

Questions

1. What effect would a higher Density altitude have on rotor RPM for the same collective position?
2. What effect would operating at maximum weight (as opposed to a minimum weight) have on autorotative RPM for the same collective position?
3. What effect would operating at a very low Density altitude (extremely cold temperatures, for example) and a very light weight have on rotor RPM and rate of descent in autorotation?

Chapter 19 Using Synthetic Training Devices to Teach Autorotations

Good Reasons for Using These Devices

There are a lot of synthetic training devices on the market (for now we'll call them all simulators, even if they don't meet that strict definition) that can be used to help teach parts of the autorotation sequence. They cannot teach all the aspects, however they cannot be used for a lot of training value at very low cost and much improved safety.

The first thing to say about using any synthetic training device, even a top-of-the-line Level D full motion flight simulator, is that the very final stages of the maneuver will most certainly be the least realistic. However, (at least in my opinion[1]), that is the least important part!

Remember that if you can get the helicopter in a position just above the ground (or water or trees) in a state of low rate of descent and low forward speed then you're likely to survive the rest of the crash.

The first thing that can be taught in a simulator with greater safety is the entry to autorotation, or more correctly, the reaction following an engine failure. The student can be shown how the rate of rotor RPM decay depends on the flight conditions and the power setting at the point of failure. The control movements that should be automatic on sensing the engine failure (simultaneous aft cyclic and down collective and appropriate bootfull of non-power pedal) should become instinctive.

A lot more simulated power failures can be put in front of a pilot in a simulator than in a real helicopter in the same amount of time.

The simulator can be used to inject engine failures in a variety of situations that would be unsafe in real life. This will add to realism and keep the student more aware of the situation they put themselves in. For example, if a student is continually making takeoffs that go into the H-V curve, engine failures can be introduced at the worst possible moment to show why this isn't a good idea. Ditto for a whole host of scenarios that would be unsafe in real life.

The fact that engine failures can also be introduced without "telegraphing" by the instructor is an added bonus. With the instructor out of view, and the ability to inject an engine failure with the push of button or click of a mouse, the student is much less prepared for an engine failure - much like the real world! Some programs even permit the failures to be pre-programmed at specific times or flight conditions, or (even better) randomly.

Once the entry has been demonstrated to a satisfactory level of competency, the next points that can be covered quickly and easily are the changes in airspeed and rotor RPM that can be used to change the flight path.

If you want to cover this during a long descent from higher altitude, the simulator can be re-positioned to this point almost instantly, instead of taking a long time in a real helicopter.

---

[1] and since I"m writing this book, it's the only opinion that matters. If you disagree- write your own book

But be careful - at least one consumer flight simulator program does not really model helicopter characteristics properly - if the helicopter is in autorotation and the airspeed is reduced to zero, the rotor RPM drops to zero (which is certainly not the case in the real world) - all that is necessary is to make sure the student (and instructors) are aware of this shortcoming.

Wind can be introduced to help demonstrate the concepts of changing flight path angle to attempt to arrive at a designated spot on the ground. Using airspeed, rotor RPM and turns, the student should be given a wide variety of situations to attempt to "hit" the necessary spot.

The constant attitude autorotation can be realistically practiced in a simulator - the methods of getting the airspeed under control and maintaining it are easily taught and carried out. The ability to judge when to apply collective is relatively easily demonstrated, as those cues are less critical than for the flare (see the next section).

The last part that can be realistically taught in a simulator is the flare. Given that a wide variety of airspeeds and rotor RPM situations can be used to attempt to arrive at a spot, the student will learn how to judge how to flare to stop the rate of descent.

The Problem with the Flare in a Simulator

Now comes the problem with using a synthetic training device that unfortunately poisons most pilots and instructors to the use of these handy devices. Most visual systems used in flight training devices do not have a lot of vertical cues in the very areas where they are needed. What are vertical cues? It's the information that comes from judging height above the ground - things like the geometric relationship between a fence post and a shrub just behind it. While a lot of these cues are generated directly in front of the pilot, a great many of them also come from vision to the side - for example, how quickly the horizon is moving up.

The cues are used in judging rate of descent with respect to the ground (this is exactly the time when the pilot should not be looking inside the cockpit!!!). For those who haven't figured it out, the transition from the end of the flare to the touchdown is exactly such a time.

Additionally, there are a lot of other cues used at this time - they are many and varied, and often not duplicated well by any device.

Aside from vertical cues and peripheral vision, another cue we use when judging the transition from the end of the flare to touchdown is proprioceptive (seat of the pants). The small and subtle changes that happen in the purely vertical plane are typically not modeled well.

The end result is that the very last part of the autorotation is not particularly realistic, and thus most often the place where pilots don't perform well. Unfortunately, this is interpreted by experienced pilots to mean that the whole experience of autorotations in a simulator is not worthwhile training. The reality is there is a lot that can be learned in even a low-cost device with good economy of both time and money as well as improved safety.

From an engineering point of view, the modeling of the aerodynamics during the flare and touchdown is difficult - there is a lot happening, and often little real-world information available for the simulator developers to go on. Just as one example, gathering the data to show with precision the thrust generated by the rotor when cushioning the touchdown compared to the rate of collective application at a range of heights within ground effect would take quite a lot flight time, often at some risk to the real helicopter.

The Big Benefit

One the most overlooked aspect of using a simulator for this training is that the flight can be played back to show the student the effects of their inputs. This can be done immediately following a particular sequence for immediate review, or it can be shown after the end of the flight. Both are useful for helping a student learn the necessary points.

A minor, but important aspect is that the computer driving the simulator sees everything, unlike us humans who have definite limitations on our ability to perceive all the things that are going on during this maneuver.

Training Instructors

Perhaps one of the most over-looked and under-utilized aspects of a simulator is the ability to train instructors. Many unusual (i.e.

unsafe or dangerous) situations can be presented to the prospective instructor in perfect safety, and with the ability to replay the scenario, the learning value is very high.

Also, it's a great place to make sure that the instructors in a school are also all teaching the same general way, using the same words and situations.

Why this isn't being exploited more is another of life's mysteries.

Summary of Chapter 19

Simulators can be a very good training aid to teach the fundamentals of many of the individual parts of the autorotation sequence, but they also have some definite limitations.

Questions

1. What parts of the autorotation sequence can be taught in a synthetic flight training device?
2. What parts of the autorotation sequence will probably not be well duplicated in a synthetic training device?
3. What are at least two benefits that can come from teaching parts of the autorotation sequence in the simulator?

Chapter 20 Autorotations at Night or in Clouds

(there's no point putting any pictures in this chapter - it's either all grey outside, or dark as the inside of a cow…).

Night Flying

For the single engine helicopter pilot, an item of serious consideration must be an engine failure at night.

Not all of us have the luxury of not flying at night. And even fewer have the option to turn down flying single engine helicopters at night if we fly professionally. So how do you manage things if the engine fails when the sun isn't shining? If you have the privilege to fly with night vision goggles, you could probably skip this chapter, as night is almost like daylight for you.

The first thing to remember about night flying is only the pilot knows it's dark. The helicopter doesn't know whether the air is clear or foggy, day or night. Only the pilot knows!

There is also no truth to the rumor that dark air has no lift..

There are many problems with piloting a helicopter at night, and they mostly relate to not being able to see things. This means it is more difficult to judge rates of closure (very important anytime, but vitally important for autorotations), It's impossible to see the wind from things like trees, grass and so on, so judging that important aspect is also impossible.

Hovering at night is made more difficult because of the lack of clarity and other cues. Shining a searchlight on the ground below the helicopter can help, but shadows change rapidly due to tilting of the fuselage and can create confusion as well. A maneuver as dynamic as the flare and landing parts of an autorotation only magnifies this problem.

Night flying is also not just going to be done around cities and airports with good lights and long, well lit runways. Hospital helipads, road accident sites, oil rigs, and so on are not normally well-lit places.

When you think about how much information is needed by the pilot to make the approach to a landing - things like rate of closure, rate of descent and so on, the difficulty of carrying out an autorotation at night becomes apparent.

One country used to insist that any single-engine helicopter without a steerable searchlight must carry long-burning flares when night flying. The theory is that in the event of an engine failure, the flares will be fired, illuminating the area below, and burn long enough for the pilot to make an autorotative approach and landing. How the pilot was to survive the ensuing grass fire, or the wrath of the local inhabitants whose houses were just burnt, was not of concern to the civil authorities. On a more serious note, the authorities had correctly realized that a non-adjustable landing light was of no use when carrying out an autorotation at night - when the helicopter flared, all the visual cues disappeared as the light pointed skywards. Something to think about.

Night vision goggles have solved much of the problem by turning night into something more closely resembling daytime, but the issues of judging height above ground for the flare are still there.

One solution is the constant attitude autorotation described earlier.

After turning into the last known surface wind, it works well at getting the helicopter to a low forward speed and low vertical speed situation with little drama, and assuming that landing light(s) work it allows the pilot to judge when to apply collective pitch in a timely manner.

In Clouds

Even multi-engine helicopters that fly in clouds can need to carry out an autorotation. Tail rotor drive failures are just one example of a reason to have to carry out a descent and landing with no engine power. The procedures are much the same as for flying at night - turn into the wind, set an airspeed of about 40 knots or whatever airspeed keeps directional stability, and wait till the radar altimeter or synthetic vision or real visual cues tell you it's time to pull up on the collective.

Summary of Chapter 20

Flying at night or in clouds presents it's own challenges. An engine failure during those conditions adds to those challenges, but there are techniques that should be known and practiced to cater for those situations.

Chapter 21   Final Words

(A Collection of Miscellaneous and Perhaps Unrelated Items.)

Inspections for Helicopters Used Consistently in Autorotation Training

If a helicopter is subjected to a lot of autorotative landings, it stands to reason that some parts are going to wear out. Things that should be looked at more closely than "normal".

Landing gear attachment points Landing gear structure, including cross-tubes and rear-mounted springs Tail Stingers - some instructors put duct tape on the stinger before flight and check it afterwards to ensure they haven't hit the stinger (more of a way of building student confidence than anything else..) Landing gear skids (underside particularly, unless the helicopter has after-market skid shoes). The aft cross-tube on those helicopters so equipped can take a dreadful beating, especially when doing landings on the grass.

Pylon mounts (I took out the spike plate on a Bell 206 when I was an instructor, and didn't realize it, even after looking directly at it!).

Where Not to Autorotate

I would try to avoid landing in the water unless you have floats (fixed or pop-out), suitable personal flotation devices for everyone on board, and training in how to get out of the helicopter when it's underwater.

Having been trained in the dunker, I can attest that the first few times you experience being underwater inside a helicopter is quite disorienting. The first time you see this should not be for real!

On the other hand, if you have no choice, or the choice is land in the water under control (for reasons other than an engine failure), then make the best of the situation.

Better Not to....

Train for autorotations on helicopters with high skid gear if you have only had experience in low-skid gear machines. The additional difference in seat of the pants can be quite alarming. If you have to, I'd suggest several landings from the hover with a count-down to make sure that you know where the ground really is!

Other Real World Issues

Wind Shear

The wind is probably not going to be constant all the way to the ground.

The friction of the ground will cause the wind speed to drop off as the helicopter gets closer to the ground (or if you prefer, the wind speed will increase as the height above ground increases). No matter which way you look at this, it means that what appeared to be just fine at 200' AGL may suddenly change at 30' AGL. If there are tall trees or buildings around, expect things to be a bit more dramatic.

Things Going Wrong

Even with good instructors and good students, things can go wrong.

Often the "going wrong" part is not caught early enough, and by the time the situation is recognized by the instructor, some drastic action is needed. Often the wrong thing to do is to try to salvage the landing, when prudence would dictate that a go-around / wave-off / overshoot is a much better option. By far the biggest error I've seen is getting too slow prior to the flare - when the flare height is reached, there isn't enough energy to stop the rate of descent. This should be easily recognizable by the instructor, but sadly appears to be missed a bit too often.

Be prepared to add power as the first step to recovering the situation - even if the engine isn't producing all it's power, some is better than none!

Instructors should know how to extricate themselves from a situation by early recognition of the problems. The ideal place to learn this is in a simulator, by the way!

Teaching Fixed Wing Pilots About Autorotations

I really don't like to make broad sweeping generalizations, however several other instructors have mentioned that sometimes trying to teach very experienced fixed wing pilots how to fly helicopters and particularly how to fly autorotations was rather challenging.

Having given "helicopter 101" to a lot of fixed wing pilots and flight test engineers, I would tend to agree, but with one caveat. High flight time in fixed wing airplanes can be a drawback, but only if it's in a small variety of airplanes. Those who have flown lots of different types adapt much more quickly and seem to understand more readily.

As a high time fixed wing pilot but a helicopter student, what do you need to be aware of? Helicopters are different! Pay attention to the exercises where you change and control rotor RPM and airspeed in autorotation - get extra proficient at these.

As an instructor what do you need to be careful of if you have a high time fixed wing student? One problem I've seen was a tendency to keep pulling back on the stick after landing! Others speak of some rather abrupt and movements of the controls in high stress situations. But you should have been aware of those tendencies in every student already.

Autorotations on Floats

Fixed Floats

If you're flying with fixed (sometimes called utility) floats, then there is a need to touchdown on the water with a speed below that which the floats can stand. The problem is that if you're flying over water it will be difficult to judge height above the water, let alone speed over the water. My recommendation is to use the "constant attitude" method in the descent, and be prepared to wait till you can see that you're really close to the water surface to apply collective. You'll already be in a slightly nose up attitude, and applying collective will slow the forward and vertical speed to an acceptable amount for a safe touchdown.

Emergency or Pop-Out Floats

The issue with pop-out floats is when to inflate them! Each system and helicopter has different instructions on how to do this, and the best advice is to know your system. Arming instructions and inflation instructions are so varied as to make it impossible to give any advice in a book like this. If you get a chance to inflate the floats during a maintenance inspection, that's a good experience, so you'll know what the noise and time to fully inflate are.

Remember the Aim of (Teaching) Autorotations

The aims of teaching autorotations are several:

increase confidence that helicopter flying is safe, develop the skills necessary for flying helicopters and most of all, learn how survive if the engine fails.

It was mentioned at the start of this book that helicopter accidents are generally more survivable than those in fixed wing airplanes because we can stop first and then land.

The number of variables that are presented to the pilot when an engine fails are quite remarkable and it's necessary to know how to use them to make the best of an unpleasant situation. The pilot also needs to know how to do this instinctively and nearly immediately.

This can only happen with training and continuous monitoring in flight.

It's too late to start making decisions after the engine fails.

Appendix A - Answers!

Chapter 1

1. Stop the drift, stop the yaw, cushion the touchdown.
2. No - keep the skids/wheels aligned with the direction of travel and cushion the touchdown.
3. To stop one skid/wheel from digging in on uneven or soft ground, causing the helicopter to roll over.
4. No! But it might be safer than hovering close to obstacles.

Chapter 2

1. The rotor RPM will increase, and the amount of increase will depend on the amount of G applied (not necessarily the bank angle, but the load factor). When the G is decreased to 1.0, the rotor RPM will return to "normal".
2. See above.
3. Most definitely the rotor will continue to turn, assuming it was turning before the speed was reduced to zero.
4. When the nose is lowered (i.e. a nose-down pitch rate), the rotor RPM will decay as the airflow through the rotor will be decreased.
5. When the nose is raised (i.e. a nose-up pitch rate), the rotor RPM will increase as the airflow through the rotor will be increased.
6. The airspeed should stay the same.

Chapter 4

1. The main symptoms of an engine failure in the cruise are: Change in engine rotor noise, Reduction in rotor RPM, A sinking feeling, A yaw to the direction of the power pedal, Perhaps a slight nose-up change in pitch attitude, Perhaps an engine failure warning light or horn and/or low rotor RPM warning light or horn.
2. Bring the cyclic back and lower the collective, then add non-power pedal.
3. The cyclic should not be moved forward until the rotor RPM is stabilized.
4. The cyclic should not be pushed forward as a first step to try to increase airspeed.
5. The symptoms of an engine failure in the hover are: Change in engine and rotor noise, Yaw in direction of power pedal, Sinking feeling, Perhaps an engine failure warning light or horn and/or low rotor RPM warning light or horn.
6. Never.
7. Rotor RPM is the most important priority. Airspeed is secondary.

Chapter 5

1. The main purpose of the flare is to stop the rate of descent. Anything else that happens is a useful side effect.
2. Reduce groundspeed and airspeed, build rotor RPM.
3. If the speed at the start of the flare is too low for the flare to stop the rate of descent, get the helicopter aligned with the direction of travel and use the collective to cushion the touchdown.
4. At the end of the flare, the helicopter should be at the pitch attitude for hovering (in an ideal world).

Chapter 6

1. The three types of energy are potential, kinetic and energy of rotation.
2. The energy of rotation.
3. The energy of rotation - without the blades turning at the proper RPM, there is no control of the helicopter.
4. No, due to the squared term in the kinetic energy equation.

Chapter 7

1. The throttle should be opened at the start of the flare and not earlier.

2. The fuel control will sense the rotor RPM (and hence power turbine RPM) increasing due to the flare and reduce the fuel flow to maintain it at the correct value. This will reduce the compressor RPM just prior the pilot demanding a large increase in power, which the fuel control cannot produce.

3. No.

4. During the descent and at the start of the flare, both student and instructor repeat the phrase - "This will be a power recovery" or words to that effect.

Chapter 8

1. To re-calibrate the various senses and provide an up-to-date measure of the performance of the helicopter.

2. Ideally yes, however if both instructor and student are proficient and current, it may be shorted.

3. To re-acquaint the senses with the personal space the helicopter is occupying as well as the location of the skids with respect to the ground.

4. No - the problem area should be sorted out first (like any education process).

5. No. Too many surprises.

Chapter 9

1. The advantages are - increased energy at the start of the flare, reduced time if in a headwind (the published max range airspeed is for zero wind), decreased glide path angle with respect to the earth.

2. Only valid for zero wind.

3. Published autorotation maximum range airspeed plus at least 50% of the headwind, up to $V_{NE\ Auto}$.

4. V_Y.

Chapter 10

1. Use maximum range technique (reduce rotor RPM to minimum, airspeed between V_Y and $V_{Max\ Range\ Auto}$), aim for spot to one side of landing area, commence turn no later than abeam landing area to zero out groundspeed at top of trees.

2. Rotor RPM.

3. Turn immediately and then sort out technique to return to spot.

4. No - you can still be in a turn up to the moment of touchdown.

5. The area that can be reached if flying into a headwind is reduced in front of the helicopter and increased to the rear (compared to a no-wind situation).

6. The tailwind can increase the distance covered in front of the helicopter, but there will still be a requirement to turn into the wind (or at least cross wind) for landing.

Chapter 11

1. Skids were designed to absorb loads on smooth, hard surfaces, the friction is less on a hard surface than on grass, there are fewer problems with differences between the direction the helicopter is pointing and the direction it's traveling.

2. Reduce the groundspeed to as close to zero as possible prior to touchdown, ensure the direction the helicopter is traveling across the ground is the same as the direction it's pointing. (you could also fit full-length skis…).

3. No.

Chapter 12

1. No - some helicopters have a pronounced difference in performance between simulated and real engine failures.

2. Report it (in writing) to the manufacturer and to the civil authorities.

Chapter 13

1. Typically, maximum weight to hover OGE at 7,000 Density Altitude, winds of less than 3 knots at the surface.

2. Takeoff power.

3. Power for level flight.

4. No - for the low hover point, the collective can only be raised following the simulated engine failure.

5. No, everyone is surprised when an engine fails.

6. In order to provide an opportunity to handle an engine failure - climbing out at a slow speed with lots of height puts the helicopter in a bad position for an engine failure.

7. For helicopters with more than 9 passenger seats, the H-V curve is in the Limitations section of the Flight Manual,.

Chapter 14

1. Not in the normal sense of the word. The rotor has no opportunity for airspace to descend through to drive the rotor.

2. Maximum weight, 7,000 Density altitude, no wind.

3. It does not appear to make much difference.

4. No - the collective may only be raised following the simulated engine failure.

5. The element of surprise at an engine failing precludes there being time to react with a down collective movement before it's necessary to raise the collective to cushion the touchdown.

Chapter 15

1. Takeoff power is applied.

2. It is anticipated that the pilot is going to have hands on the collective (as well as the cyclic and feet on the pedals).

Chapter 16

1. Not if you don't have to.

2. Even worse idea than hovering at the high hover point, unless you can't avoid it.

3. There will not be sufficient time or altitude to accelerate to an airspeed that will allow a flare and landing that will not damage the helicopter.

4. You should apply a substantial margin to the published H-V curve.

Chapter 17

1. There are a great number of different autorotation techniques that can be used. The airspeed for a constant attitude autorotation in a 5 knot wind should be around 40 KIAS.

2. The ground rush.

3. Range and rotor RPM would be used to arrive at a location that would put the helicopter in a low-energy state within the boundaries of the back yard.

Chapter 18

1. At a higher density altitude with the same collective position (and all other things being equal) the rotor RPM will be higher.

2. The higher weight would have a higher rotor RPM for the same collective position (all other things being equal).

3. The rotor RPM will be significantly lower in the cold weather and low weight condition.

Chapter 19

1. The entry, descent (including changing rotor RPM and airspeed) and the flare.

2. The portion from the end of the flare to touchdown.

3. Increased safety and a much less steep learning curve.

www.ingramcontent.com/pod-product-compliance
Lightning Source LLC
Chambersburg PA
CBHW051225200326

41519CB00025B/7249